SKETCH

OF THE

MEDICAL TOPOGRAPHY,

OR

CLIMATE AND SOILS,

OF

BENGAL AND THE N. W. PROVINCES

SKETCH
OF THE
MEDICAL TOPOGRAPHY,
OR
CLIMATE AND SOILS,
OF
BENGAL AND THE N. W. PROVINCES

BY

JOHN M'CLELLAND

GYAN

Published by

GYAN

Gyan Publishing House
5, Ansari Road
Daryaganj, New Delhi-110002
Phone: 011-47034999, 9811692060
E-mail: books@gyanbooks.com

Distribution Network
gyanbooks.com
India, USA, Canada, UK, Australia

ISBN: 978-93-6433-029-9 (HB)
First Published, 1859

2nd Impression 2024

Printed at: Gyan Press, Delhi.

SKETCH OF THE MEDICAL TOPOGRAPHY, OR CLIMATE AND SOILS, OF BENGAL AND THE N. W. PROVINCES

Author: JOHN M'CLELLAND

SKETCH

OF THE

MEDICAL TOPOGRAPHY,

OR

CLIMATE AND SOILS,

OF

BENGAL AND THE N. W. PROVINCES.

BY

JOHN M'CLELLAND, F.L.S., F.G.S.,

SURGEON, BENGAL SERVICE; CORRESPONDING MEMBER OF THE ROYAL ACADEMY OF SCIENCES, TURIN, OF THE ROYAL BOTANICAL SOCIETY, RATISBON; ETC., ETC., ETC.

LONDON:

JOHN CHURCHILL, NEW BURLINGTON STREET.

MDCCCLIX.

PREFACE.

It was my duty to make some communication on the circumstances connected with the prevalence of cretinism in the plains of Bengal, with which I became acquainted in the early part of last year. It was necessary, in doing this, to refer to facts previously elicited on the same subject in Kemaon. The work in which those results originally appeared in India being out of print, and but few copies of it having reached England, the circumstances detailed were consequently imperfectly known in this country, and that only through the medium of reviews. For these reasons, as well as the great interest of the subject, I determined on giving the result of inquiries in Kemaon in full, as they originally appeared. In doing this, I have taken the opportunity of offering a few brief observations on the climate and soils of Bengal, and the N. W. Provinces generally.

On the subject of temperature, and the fall of rain, much assistance has been derived from the official registers kept, at 127 stations, by the medical officers of the Bengal Presidency, placed by Dr. Lamb, late Physician-General, in the hands of Colonel Sykes, M.P., by whom they were submitted, with valuable remarks of his own, to the British Association of 1851, in whose Reports of that year they will be found.

By grouping the stations according to the physical features of the country, the plains—generally supposed to present but one climate, with vague undefined distinctions between the Upper and Lower Provinces—are shown, on the contrary, to possess numerous distinct climates, each having its own annual temperature, fall of rain, and prevailing winds, points that have never been fully or sufficiently considered with reference to their effects on the human constitution, or even in the selection of stations for European troops.

Although we have little information regarding the climate of Lucknow, or experience of its effects on European health, to guide us, yet, according to the little we do know, as well as the principles developed in the following pages, it will be found to range with the climates of the Gangetic plains, and more especially with that of Cawnpore—the most fatal in the N. W. Provinces.

On the other hand, Fyzabad, and the stations beyond the Gogra, will be found to present a climate more congenial to the European constitution, ranging with the more temperate climates of the eastern plains, to which they belong. I give the more prominent expression to this timely statement, that it may be of use in regulating the fixed European establishments in Oude, consistently with the sanitary views which may be derived from the following pages.

In the work of Dr. Martin, *On the Diseases of Tropical Climates*, the influence of soils is referred to in a manner calculated to draw the attention of medical officers, more especially in India, to this important branch of inquiry. I have, in the following pages, endeavoured to sketch out, though faintly and imperfectly no doubt, the distribution of soils in Bengal and the N. W. Provinces, with a view to the general application of that question to the investigation of the endemic cause of disease.

I have also offered a few remarks touching the influence of soils, rather with a view of eliciting information and suggesting inquiry, than from anything of importance on the subject I had to communicate, beyond what relates to cretinism. As regards this subject, the results are not perhaps altogether unimportant; and as they are new, as far as relates to the prevalence of the disorder in the plains, I may mention, that in cretinism the development of the mind becomes suspended in childhood, and imbecility is combined with the irregular growth and consequent deformity of the body, often attended with want of power of speech and sense of hearing.

In the districts where the malady is endemic, persons thus affected are called *Bowkuts*, the common term for idiots in Bengal. The Bowkuts of Goruckpore and adjacent districts on the borders of the Tarai, are the children of persons subject to goitre, a disease which is shown to be the result of continued residence on a certain kind of soil.

A large proportion of the children born in certain villages in these parts are subject from an early age to this disease, which, from the knowledge we now possess, may undoubtedly be arrested by a change of residence from one soil to another, especially when accompanied with a corresponding change in the topography of the wells from which the accustomed drinking water is taken.

On the subject of cretinism, a most important fact was communicated by Professor Kölliker to the Swiss Association, at Basle, in 1856—namely, the early ossification and subsequent thickening of the base of the cranium in cretins, attended with the obliteration or narrowing of the foramina. The communication, however, appears to have been made with a view to the means of instituting more comprehensive inquiries, rather than of announcing ascertained results, and was naturally addressed to residents in places where goitre and cretinism prevail.

As mortality is great amongst the cretin population of the districts above alluded to, crania may be there found, calculated materially to aid this important inquiry.

It may, however, be remarked, that the influence of soils is not confined to one form of disease, or to one class of persons, but applies equally to the British soldier and all who are exposed more especially to the influence of tropical climates. Investigation on the subject is naturally attended with many difficulties, more particularly as regards malaria.

The object of our search, in the case of jungle fever, leaves no living monuments to point out the localities in which it prevails at certain seasons in all its intensity. These localities are either known only to be avoided, or they are discovered by accidents which take us by surprise and leave no time for deliberate investigation.

On the other hand, the effects of malaria, so far as they usually fall under the observation of physicians, are seldom free from numerous complications, which render it doubtful how much is due to malaria and how much to other causes arising from epidemic influence, or the effects of high temperature, exposure to the sun, intemperance, fatigue, and the like; thus obscuring the question of malaria.

Notwithstanding these difficulties, we may hope for much from the general progress of science, and the application of its results to the investigation of endemic disease.

LONDON, 1859.

CONTENTS.

SKETCH OF

THE MEDICAL TOPOGRAPHY

OF

BENGAL AND THE N. W. PROVINCES,

&c. &c.

GENERAL DESCRIPTION OF THE MOUNTAIN PROVINCE OF KEMAON.

BURMDEO, one of the passes into the mountains of Kemaon, is formed by the defile of the Gogra river, where it breaks through the outer range of mountains to enter the plains. The pass, or narrow gorge, is characterised by broken lofty cliffs, barely sufficient to allow of the escape of the torrent; so that the path takes a winding turn, in which the view of the river is lost soon after entering the hills. The ascent lies, for the most part, along narrow ravines and undulating ridges, enclosed within lofty precipices, until the crest of the first high ridge of mountains is attained with considerable toil.

After having gained the summit, a still higher range of mountains in front is found to intercept the view of the Himalayas, and the path descends rapidly to the encamping ground, which is situated in a deep narrow river valley, without inhabitants, but clothed with rich tropical vegetation.

The malaria of the place is such that this pass is only available, and that for travellers on foot, from November to March. Indeed, independent of any other cause, vegetation is so dense and of such rapid growth, that it would be very difficult, where there is so little traffic, to preserve an open communication by this route.

The second day's journey from the Ludhoo river, at Belket, is one continued ascent to Chourapany, for the fatigue of which the traveller is rewarded by the first view of the snowy range. From Chourapany itself, the view is rather circumscribed; but on ascending one of the neighbouring heights, an uninterrupted panorama of slender summits is seen to extend, like a vast white curtain, from the north-west to the east. Considering the distance of the nearest snowy peak to be not less than 60 miles, the extent of the chain embraced at one view from this point cannot be less than 100 miles. The mountains intervening between the snowy range and the eye, and upon which you appear to look down, vary in height from 6,000 to 10,000 feet. The different branches of the River Gogra—or Kalee, as it is here called—are seen in some places and only heard in others, rushing in torrents through the valleys that divide the mountain groups from each other. The great valley of the Kalee itself, which separates the province of Kemaon from Nepaul, is seen a few miles to the east.

While the mean height of the river valleys may

be about 1,500 feet above the sea, the height of the adjacent mountains rising from them varies from 3,000 to 12,000 feet. The mountains are massive, and, although generally covered with forests, still preserve the characteristic outlines derived from the rocks of which they are composed. The highest continuous ridges are generally formed of granite, succeeded by mica slate, gneiss, and clay slate.

At lower elevations, hornblende is chiefly found, together with greenstone and other similar old trap rocks. Clay slate is largely developed between the elevations of 4,000 and 8,000 feet, resting on one or other of the rocks already named. In some places, dolomite passing into green sandstone, in others, compact blue limestone, both resting on clay slate, form equally bold and precipitous mountains.

Hornblende slate forms steep and rugged, though well-wooded, declivities. Mica slate, gneiss, and clay slate afford more gentle slopes. The limestone mountains are abrupt and precipitous, with deep narrow valleys. The green sandstone, like the limestone, forms lofty steep mountains, which are seldom wooded to any great extent; while ravines and low places, situated at the base, are strewed with masses which have been precipitated by decay from above.

The portion of Kemaon examined, is embraced by 29° to 29° 45′ N. lat., and 79° 55′ to 80° 20′ E. long., and lies on the western side of the Kalee.

The valleys formed by the great tributaries of the Kalee divide the district into sections.

The first is that deserted tract already noticed, lying between Belket and Burmdeo Pass, crossed on entering the mountains. It is entirely composed of calcareous and argillaceous grit.

The second is a more important section, and extends from the River Luddoo, at Belket, to the Ramessa river, or great western branch of the Kalee. A ridge of granite composes the centre of this section, and forms occasional elevations of nearly 8,000 feet. Gneiss, hornblende slate, mica slate, and clay slate, together with green sandstone and dolomite, are the other rocks composing this section.

The third natural division is that which lies on the north of the Ramessa river, and between the Kalee and Surju, embracing some cultivated, though small valleys, more especially the valley of Shore, which gives its name often to the neighbouring district.

Of the mountain rocks composing the Shore district, clay slate forms the basis on which the others rest, while it ascends to elevations of 8,000 feet.

Blue compact, or what is called alpine limestone, resting on the clay slate, gives a peculiar character to the mountains of this neighbourhood. Each mountain, standing almost detached from the group to which it belongs, is usually distinguished by a particular name derived from a temple on the summit. "In the infancy of civilisation," says Humboldt, "the high places were chosen by the people to offer sacrifices to their gods; the first temples, and the first altars, were erected on mountains."

The same practice prevailed at one time in Burmah, where the ruins of a temple may be still seen on the summit of each mountain you pass, in ascending the Irawaddi from Prome. The inhabitants of the Khasia hills select the summits of the highest mountains for the purpose of burning their dead; so that the veneration of high places would seem to have been naturally ingrafted in the mind of man; and it is difficult, even in the present day, altogether to disregard this feeling.

ROCKS OF KEMAON.

Granite.—This rock is found at Chourapany, penetrating through gneiss, and forming a succession of elongated ridges, which, if not the loftiest, at least constitute the bases of the highest districts in Kemaon.

The ridge alluded to extends in a north-westerly direction for 40 miles, when, with increasing altitudes, it joins the snowy range or high mountain chain, after forming the Leti, Tirsal, and Dhanapur mountains on the east side of the valley of the Alacunanda river. A great basin, composed of granite, is thus formed, which is filled up of mountain groups composed of other rocks. The granite, as has been stated, makes its appearance only in the centre of the ridge in the loftiest places, where it occasionally appears to be stratified, the strata taking the direction of the range, and apparently nearly vertical. In other places, it would seem to consist of round masses, subject to concentric disintegration, without evidence of stratification. It is of a grey, or sometimes a reddish colour, but generally bluish-grey. The quartz is crystalline; the felspar dull, earthy, and fine grained —specific gravity, 2·71.

Gneiss, reposes on the granite in conformable strata,

and the transition between the two rocks is by imperceptible degrees, so that it is difficult to say where gneiss begins and granite ends. In the newer granite, the quartz becomes less crystalline, and of smaller quantity in proportion to the other ingredients, until it disappears, leaving chiefly, felspar and mica, with a very small proportion of quartz. The change renders the rock less compact, and in this state it occurs at the base and on the slopes of Chourapany; and from thence it extends in a north-westerly course, forming the principal portion of the most elevated district in Kemaon.

Its course is marked by immense denuded masses of granite as well as greenstone, which usually follow the line of junction with granite. Sometimes these masses are grouped together in the form of cones and pyramids; at other times, enormous masses of rounded form are loosely accumulated on the verge of a precipice, where they are so nicely balanced that the least force would be thought sufficient to precipitate them to a fearful depth. Round detached overlying masses of this sort are very common in all the granitic districts, which compose large tracts of the plains, particularly in Southern India.

In tropical climates, perhaps gneiss is more subject to decay from exposure to high temperature than in northern climates; a curious illustration of which may be adduced by the ruins at Kalee Kemaon, of an old city. It was erected on gneiss, and appears to have been totally destroyed by a *canker* in the rock on which it stood. A few vestiges only remain—part of a citadel, a balcony, a solitary door-way, and, here and there, the remains of a fountain and the ruins of temples, all in carved granite—to attest the former importance of the place.

Hornblende Slate.—The lower parts or bottom, as it were, of the great granitic basin already alluded to, are chiefly filled up with this rock. The bed of the Ramessa river, only 1,500 feet above the sea, is formed of it, and it forms the troughs of most of the deep narrow valleys, through which the rivers ramify between the mountains, seldom rising to elevations above 2,000 or 3,000 feet. The high temperature and miasmata of such places would alone account for their being uninhabited, at least, for the greater part of the year. Besides which, the rugged character of hornblende slate, the tabular strata standing several feet above the surface, renders cultivation impossible, while it promotes the growth of indigenous forests.

It is composed chiefly of hornblende, having flattened laminæ and grains of quartz imbedded in alternate streaks, which give the rock a slaty character, and render it somewhat sectile.

It passes, on the one hand, into gneiss, on which it rests, and on the other, into mica slate.

This same rock has been found in Central India, particularly in the Rajmahal hills, underlying altered coal measures in connexion with other trap rocks, as well as in the Bhagulpore district, in connection with syenite, quartz rock, and gneiss.*

In some places it contains extensive beds of

* *Report of the Geol. Survey of India*, 1848-49, p. 33, tab. 7 and 9.

micaceous iron ore, as in the Ponar valley, near Lohooghat.

Mica Slate.—This is generally found resting on hornblende slate, or where that is wanting, as in the higher levels, on gneiss.

It is composed of small crystalline grains of quartz and felspar, with a considerable proportion of mica. Sometimes the quartz and felspar are wanting, while the rock is composed of clay slate, and mica; on such occasions, this clayey form of mica slate, alternates with extensive beds of quartz rock. Specific gravity, 2·6.

Mica slate is also found extensively in Central, and other parts of India, particularly in the Beerbhoom district in Bengal, where it contains ores of copper, at Baragundah.*

Clay Slate.—This rock composes one sixth part at least of the whole superficial surface of the province, and is stretched in conformable strata over mica slate; and, where that is wanting, over hornblende slate—both of which rocks are frequently found to pass into clay slate by regular gradations. It rises into sloping ridges of 7,000 feet high, having rich and fertile valleys, well peopled, and affording a pleasing contrast to the inhospitable character of those districts which are composed of hornblende slate.

This rock is soft and friable, as its tenacity depends on ferruginous parts easily decomposed by the heat and moisture of the atmosphere, these causes keep up a constant and uniform process of decay, which

* *Reports of the Geol. Survey of India*, 1848-49, p. 23.

affords everywhere an abundant soil on the slopes, and gives to the outlines of mountains composed of clay slate, a smooth softened appearance. The ferruginous clayey soil thus formed, though sufficient, from its retention of moisture for the growth of dense vegetation, is often carried down into the valleys over beds of limestone and marl; and, by the calcareous admixture thus acquired, the ferruginous oxides are decomposed, and a soil afforded capable of yielding, in these valleys, with little labour, several crops annually.*

In its most characteristic form, clay slate is represented by what is familiarly known as "roofing slate." It is only certain beds, however, that afford this valuable material, which are generally known by their fine slaty cleavage. In its more common form, clay slate is of a dark-greenish grey colour, with a laminated dark earthy structure. It is very distinctly stratified, and the strata are intersected by seams passing at right angles to the seams of stratification.

It contains veins of quartz, but no minerals of any value, and it seldom occurs sufficiently sectile to afford roofing slate. The specific gravity of the old blue variety is sometimes as much as 2·6, while that of the greenish-grey variety, which has a loose friable texture, is as low as 2·3.

This rock occurs to a considerable extent in the Coruckpore hills, where beds of old blue state, identical in all respects with the Welsh roofing slate, are found at the pass between Bhem Bham and Goorma.†

* "Are any of the salts of iron present? They may be decomposed by lime."—*Davy's Agricultural Chemist.* What the philosopher recommended to be done by art is here performed by nature.

† *Report of the Geo. Survey of India*, 1848–49.

Alpine Limestone.—This rock is largely developed in the neighbourhood of Shore valley, where it forms bold, lofty, precipitous mountains of itself; or, resting on low ridges or on the sides of mountains composed of clay slate, forms bold mountain tops and shields. It is a compact smoke grey-coloured limestone, presenting, in the great scale, an appearance of lines of stratification, but which become obscure and doubtful on closer examination. Where it rests in mantle-shaped beds on the sides of clay-slate mountains, it forms precipices and assumes a slaty structure from layers of foreign matter intermixed with it. In such circumstances it is less pure, leaving a considerable residuum undissolved in muriatic acid. Where it forms mountains of itself, it dissolves entirely in that acid, and assumes a massive brecciated structure incrusted with stalactites. At the foot of the precipices, knolls and small hills are formed, composed of masses reunited by calcspar; and, at still lower levels, small hills and banks of conglomerate occur, composed of rolled masses of limestone, clay slate, quartz pebbles, and other rocks, cemented together by calc tuff.

This last deposit becomes very common in the Shore district, and wherever there are mountains of limestone.

In other situations in the same district, beds of fine granular limestone occur, varying from snow white to peach-blossom red, without lustre.

A black variety of the compact rock also occurs in overlying isolated masses on clay slate in the cantonment of Petorahgurh in Shore valley.

These varieties of limestone dissolve rapidly, with effervescence, in muriatic acid, and present a specific gravity of about 2·6.

In the valley of Barabice, ores of yellow and grey sulphurets of copper are found in the lower beds, of the blue slaty variety, in which these ores are disseminated, along with talc, in strata seams and rents. The slaty layers of the limestone vary from half an inch to several inches in thickness, and consist of blue compact limestone. These layers or tabular planes occur broken into fragments, usually from a foot to six inches square. It is in the interstices of these fragments that the copper ores are found.

In the mines, which are long and narrow excavations, penetrating from 50 to 100 yards and more into the rock, lying masses of talc are seen interposed between the limestone strata. The richest repositories of the ore are situated at the junction between the limestone and clay slate, in beds of dark bituminous talc.

The mines are farmed to natives, and worked, with little profit, by cretins, most of whom are deformed either with goitre or with large heads and dwarfish frames, with little or no mental faculty.

Green Sandstone.—This rock rests on clay slate. Its colour is greenish or yellowish grey, often tarnished with red or brown delineations on its surface. Structure, fine granular, presenting a large slaty cleavage. Fracture of fresh specimens, splintery; when fresh, it is hard enough to scratch glass, but it becomes softer and assumes a dull, earthy appearance, on exposure. Specific gravity, 2·6.

This rock forms lofty conical peaks and elongated sloping ridges, often very steep, but seldom mural precipices. The Sooee mountain, near Lohooghat, is composed of it; from thence it extends 40 miles to the westward, and is found on the ascent leading to Almorah, from Lohooghat.

It passes into dolomite and may be a variety of that rock, although it appears to me to consist of fine silicious grains imbedded in a talcose matrix, which changes, on exposure, from a close-grained almost compact rock, to a granular sandstone.

It is not confined to Kemaon, but is found also in Hindostan, in the Monghyr district, where it forms a ridge of hills at Guidore.

Compact Dolomite.—Low ridges composed of this rock are seen rising abruptly out of the narrow lateral valleys at Belket. The beautiful green and yellow colour of their naked precipices, and picturesque form of their inaccessible summits, as well as the uniform arrangement of their massive and nearly perpendicular strata, all combine in producing a striking effect.

This rock rests on hornblende slate, or is separated from it by a curved slaty combination of quartz and chlorite, it passes into granular quartz, and this last into dolomite. The strata are nearly vertical, or are inclined at an angle of 75°, being divided at right angles with the planes of stratification, by numerous rifts and fractures, occasioned by the encroachment of torrents on the base of the mountains, which thus undermine the strata, causing its displacement. The percolation of water through the strata seams, also tends to the same effect, as well as causing land-slips.

The compact dolomite of this locality may be described as a of greenish-grey colour, passing, on the one hand, into somewhat of an emerald green, and, on the other, into azure blue. The lustre is pearly, inclining to vitreous in the fresh fracture; fragments, sharp-edged, and hard enough to scratch glass, translucent on the edges and rather brittle, rather meagre or talcose to the feel. Specific gravity, 2·65.

Granular variety.—At the bridge over the Ramgungah river, on the road between Petorahgurh and Almorah, 40 miles to the north of the first-mentioned locality, the same rock again occurs. The granular variety here constitutes the bed of the river; it ascends on the left bank, and is found in the lower parts of the Gorou valley, 2,000 feet above the situation it occupies at the bridge.

On the opposite or Gungowly side of the river, the rock assumes a coarse granular and loosely aggregated form, consisting of ovate grains, as it would seem, of quartz, imbedded in a talcose matrix, the whole presenting a greenish colour and pearly lustre.

It rests on clay slate, and forms a lofty ridge of mountains, stretching to the north-west, between the Ramgungah and Surju rivers.

The ridge is narrow at its crest, and extremely steep; its outline is diversified by numerous bold conical peaks, one of which—the Rye Peak—being upwards of 8,000 feet high.

These rocks contain no mineral beds of any value.

Calcareous Gritstone.—This is found composing ridges near the entrance of the pass. It is composed of fine sandy particles imbedded in a calcareous cement, the whole presenting a greenish tinge, derived from the prevalence of hornblende and green sandy particles. It has an earthy appearance, and effervesces strongly in acids when about a third part is dissolved.

This rock, or a substance possessing the same mineral and chemical characters, is found in the plains of the Doab underlying the beds of kunkur, where, in the Jumna, between Allahabad and Agra, under the name of flag, it has been found obstructing the navigation of the stream. A similar rock, occupying the same position, has been found in the bed of the Gogra river, also underlying beds of calcareous clay or kunkur; so that it is probable that this rock extends largely beneath the beds of calcareous clay and river deposits which form the plains of the N. W. Provinces, forming the basis on which the kunkur and river sediments rest.

Slate-clay.—This is a still more superficial rock, resting upon the calcareous grit, and forming low banks and hills at the foot of the mountains, covered irregularly by small hills of boulders.

It is a sandy, dark, soft rock, rendered slaty by the large proportion of mica it contains. It is frequently found at the foot of mountains; and occurs forming a low hilly tract, at the foot of the Garrow hills in Bengal. It is also found extensively in Pegu. In both these last cases, it is found immediately under the laterite; and in Pegu, it often assumes a basaltic character, in connexion with trap rocks.

Alluvial Deposits.—In mountains the alluvial deposits are much more simple than in the plains, because each rock is covered by the alluvium derived from its own disintegration.

They are silicious, when reposing on granite. The quartz contained in that rock is pure silicious earth, and the same substance combined with iron, is contained in the felspar and mica, which decomposing, liberates the grains of quartz, and the whole forms a yellow clay, the colour being derived from the oxidation of the iron.

They are aluminous, when derived from the disintegration of clay slate and trap rocks, both of which owe their hardness and strength to the ferruginous parts which enter into their structure; these, decomposing, form yellow oxides, which give their colour to the clay.

The foregoing are sedimentary mixtures of insoluble matters. There is, however, another class of deposits, which owe their character to matters held in solution by water, of which it is only necessary to mention one — calc tuff. This is spread as a cementing matter over other materials, thus forming a solid conglomerate, called in Switzerland "nagelplugh." We find this in Kemaon in the neighbourhood of limestone mountains, not only in valleys, and ravines at the foot of precipices, but at considerable elevations, where the fragments of which it is composed are small, becoming larger and more rounded as they descend into lower places.

It is even seen in the course of formation at springs in alpine limestone, from which the waters issue with ebullition, occasioned by the escape of carbonic acid; and as this takes place, lime is deposited from the water, encrusting all objects exposed to its action, and consolidating the sand and gravel in the vicinity of the spring.

TABLE-LANDS AND PLAINS.

BENGAL AND THE EASTERN PROVINCES.

THE foregoing brief sketch of Kemaon, its rocks and valleys, may facilitate more extended inquiries on the subject in other provinces, without the necessity of going into such minute details.

The Khasyah mountains form the north-western extremity of the group which separates Assam from Bengal. The elevation of the chief ridge is here 6,000 feet high, presenting broad table-lands at an elevation of 5,000 feet.

Cherraponjee is situated on this range, 500 feet lower than the table-land; and although the fall of rain at this place is excessive, yet, the soil being dry and rocky, as well as inclined, the elevation is highly approved of by those who, from long residence and other experience, are best competent to give an opinion in such matters.

The Khasyah mountains terminate in the Garrow hills at the bend of the Burhampooter, where it enters Bengal from the Assam valley. On the south, the range is prolonged, under various names derived from its several aboriginal races of inhabitants, to Aracan, forming a mountainous coast to Cape Negrais.

From Aracan, the range extends to the northern parts of Burmah, after forming the hilly tract between that kingdom and Upper Assam. The elevated valley of Moneypore is situated in the centre of this group.

As far as the rocks composing it have been explored, the group consists of gneiss, clay slate, sandstone, and coal measures, the latter reposing in small outlying patches on the flanks of the range on the Bengal, as well as the Assam side, in Cachar, and on the coast of Aracan.

The highest elevations that have been correctly ascertained, are from 5,000 to 6,000 feet.

The Yomah hills form low ridges composed of trap, argillaceous schists, and sandstone, intermixed with an earthy limestone, all resting on gneiss, which separate the valleys of the Irawaddi and Sitang rivers, in the upper parts of Pegu.

On the east and south of the Sitang, the Shan mountains form more lofty ridges, composed of gneiss and other crystalline rocks. Taking a southerly course, these ranges form the centre of the Malayan peninsula. Coal measures, with limestone, rest on both sides of this chain in the Shan states, as well as in the Tenaserim provinces.

Laterite, or red clay, is found in all these provinces, filling up depressions and forming plains and vallies, underlying the recent sandy sediments of the rivers and estuaries.

This laterite is the only rock, if we may so term it, of general occurrence that has not been already described. It is a dense heavy clay, usually of reddish colour, passing into dark brown and grey, often variegated red and white—the white parts consisting sometimes of vesicles filled with white magnesian clay, quartz pebbles, or white fine sand, and very often with small round nuclei of oxide of iron. Exposed, as it often is, to the action of water in the beds of rivers and on the coasts, it is soft and may be dug out in the form of a brick, which on drying becomes a hard, durable, building material. When composing the surface uncovered by lighter soil, it forms heavy marshy plains, abounding in jeels or lagoons during the rainy season, which, drying up by evaporation, leave the surface cracked and rent by open fissures in the form of a tesselated pavement, as in the lower parts of Bundelcund, and of the Therawaddi district in Pegu, as well as in Sylhet. In hilly districts it forms a barren arid soil, incapable of cultivation, overrun with stunted trees and other jungle. When covered with a lighter soil, it affords rich and fertile land.

At Jumalpore and Mymensing in Eastern Bengal, this laterite clay forms the bed of the Burhampooter river, as well as the low marshy plains extending to Sylhet, and along the left bank of the Burhampooter into Assam, where it underlies the river deposits of that province, resting occasionally on slate-clay, trap rocks, and gneiss.

From Assam and the eastern parts of Bengal, the laterite is traced along the eastern shores of the bay to Amherst, and other parts of the Tenaserim provinces. From thence it extends into Pegu, along the valley of the Sitang river to Tounghoo, where it underlies the river sediment, and compases the hilly tracts and plains extending from the right bank of the Sitang to the valley of the Pegu river, and the extensive plains from thence to Rangoon. It also forms the marshy plains in the Therawaddi and Henzada districts in the valley of the Irawaddi, and is found throughout Pegu, reposing on dark argillaceous slate or trap; and, where these are wanting, it rests on gneiss. In the lower portions of the plains and valleys of Pegu, it is covered with river sediment, and south of Rangoon with the sediment of the Delta.

While the term laterite is here applied in general to clay, it may be necessary to explain why it is held to be different from sedimentary deposits. Drs. Voysey, Christy, Cole, and Carter, as well as other distinguished observers, consider it to be of volcanic origin. Drs. Buchanan Hamilton, Clark, Benza, and others, consider it to be the mere result of the decay of granite, clay slate, and trap. It is very probable, therefore, that various clays, derived from different sources, are here confounded under the name of laterite. They are all, however, highly absorbent and retentive of moisture, and, when uncovered by a more porous superficial soil, exercise a considerable influence on tropical climates, for which reason they are here classed under the same head.

Laterite, strictly so called, varies in different places, and in different beds in the same place. The lower beds often contain a larger proportion of iron in the form of small nodular concretions.

The laterite forming the banks of the Burhampooter in Assam and Eastern Bengal, as well as on the coasts of Ceylon, is variegated in colour, containing small cavities filled with white sand, and often, especially in the table-lands of Nagpore and other places, the cavities are filled with green earth. For an excellent general description of laterite, and summary of various opinions on the subject, I must refer to a paper by Dr. Carter, in the *Journal of the Bombay Branch of the Royal Asiatic Society* for July, 1852.

BENGAL AND THE N. W. PROVINCES.

Turning our attention now, from the south-eastern frontier, and Tenaserim provinces, to Central India, we find this region composed of small but distinct mountain groups, the highest rising to an elevation of about 5,000 feet, under 81° 50′ E. long., and between 22° 40′ N. lat. at Amarakantak, in Chota Nagpore, and at Patchmaree, in the Nagpore district.

The higher Hills and Table-Lands.—The northern portion of the group comprising the Aravully range consists entirely of granite, gneiss, syenite, mica slate, and clay slate, each appearing in its turn, composing elevations of 3,500 feet under 74° E. long. and 25° N. lat., diminishing in altitude until the range is lost in the plains of Delhi and Hansi.

The southern boundary of the table-land is formed by the Vindhya range, consisting chiefly of gneiss and sandstone, with beds of trap and altered coal measures. It is joined to the Western Ghauts by means of the Sautpoora range, which forms a hilly tract on the south side of the Nerbudda valley, where it rises to an elevation of 4,800 feet at Patchmaree, forming a table-land under 78° 15′ E. long., and 22° 30′ N. lat. The Vindhya range extends to the east under 23° N. lat., forming elevated table-lands at Sohagepore, which extend from Amarakantak half-way to Jubbulpore. It is here where the sources of the Sone and Nerbudda rivers originate. On the east, the Vindhya range expands into the table-lands of the south-western frontier, Nagpore, and the Saugur and Nerbudda territories, bounded by the higher valleys of the Sone, the Baracka, Damooda, and Mahanudee rivers, as well as by those of the Chumbul river on the north. These table-lands are chiefly composed of gneiss and sandstone, flanked at intervals by coal measures, found in the valleys of some of the above-mentioned rivers and their tributaries.

The Vindhya and Aravully ranges are connected at their western extremity by the hills of Rath, forming the eastern boundary of Guzerat, and the western passes of the table-land of Malwa, composed chiefly of sandstone and trap rocks, which form the north side of the Nerbudda valley and the upper part of the descent into Guzerat—the lower portion of the descent to Guzerat, as well as the plains along the eastern border of the table-lands on the Bombay side, being formed, for a breadth of 60 miles, of granite.

The highest portion of the table-land extends through the centre of Malwa, from Ougein, in the direction of Seronge, lying on the north-west of Bhopal, about 2,000 feet high, although the elevation of this portion of the table-land has not been ascertained correctly.

On the east, the table-lands of Malwa are bounded by the ghauts and low hilly province of Bundelcund, forming the south bank of the Chumbul river, and embracing the country from thence to the Sone, composed of sandstone and trap rocks resting on granite, with a small, detached ridge, composed of sandstone and beds of limestone, called the Kaimur range.

The north side of the Chumbul valley consists of the Chittore hills, a small range composed of sandstone and quartz rock, extending on the east nearly to Agra, diverging on the north through Ajmere, and uniting in that direction with the Aravully range, thus enclosing the valley of Mewar, with Odepore at its western and Ajmere at its eastern extremity.

The Santal hills form a small mountain group on the east side of the Baracka valley, which separates this group from the table-lands of Hazareebagh. These hills extend to Rajmahal, and Monghyr, on the Ganges, varying in altitude from 1,500 to 2,000 feet, and are composed of gneiss, mica slate, and clay slate, sandstone passing into quartz, and trap rocks, with productive as well as altered and barren coal measures in the valleys.

The table-lands of Central India thus consist of rugged hilly tracts, sometimes presenting moderately-elevated plains intersected by broad valleys, often forming the beds of rivers.

The valleys are not generally unhealthy, although presenting a higher temperature than the open plains of India, while some of the elevations are sufficient to afford very marked improvement of climate.

The laterite which forms the marshy plains of Sylhet, disappears beneath the deposits of Bengal. It reappears in the plains at Midnapore, and from thence stretches over the southern peninsula of Hindostan. It extends to the north as far as Neemuch, and west, through the valleys of the table-lands and along the southern flanks of the Vindhya and Sautpoora ranges to Bombay, where it presents the same characters as it does in the plains of Sylhet and Pegu.

In the vicinity of the trap rocks of the Deccan, Malwa, Bundelcund, and Rajmahal, the yellow clay, when it appears at the surface, assumes a dark colour, and passes into a black soil. The same thing is observed in the yellow clay formed from the decomposition of trap in the hills between Prome and Tounghoo, in Pegu.

From Bundelcund, the dark clay, changing to a lighter colour, passes under the deposits of the Doab; here it is impregnated with calcareous matter, under the name of kunkur, and is found resting on a micaceous sandstone flag in the bed of the Jumna between Allahabad and Agra, where it is buried to a depth of from 80 to 150 feet beneath the light sandy deposits of the Doab.

At Murnaghat, about eight miles below Fyzabad, in the bed of the Gogra, a grey-coloured slaty rock, consisting of argillaceous matter, mica, silex, and lime, is found when the water is very low in the dry season, precisely similar to the sandstone flag of the bed of the Jumna.

Upon it rests a bed of laterite, soft and plastic when exposed to the action of the water, but becoming hard on exposure to the air, with a rough pitted surface, in all respects corresponding in its characters with the laterite of Sylhet and Pegu, except in containing some portion of lime, which is wanting in the laterite clays of Southern India.

At Gorughat, 30 miles lower down, near Tanda, the bed of the Gogra is partially obstructed by ledges of the same slaty rock as that above noticed at Murnaghat, and the clay, containing nodules of calcareous matter, is brought much nearer the surface, being buried beneath a thinner covering of sand.

Several other observations, in various places in the bed of the Raptee as well as in that of the Great and Little Gunduk rivers, tend to show that in the districts east of the Gogra river, in Goruckpore, Champarın, and Tirhoot, the calcareous clay is far more lightly covered than in the Doab, and, indeed, forms considerable tracts of the surface soil, under the term of "bhat lands," in these districts.

Sedimentary Deposits.—The plains of Bengal and the N. W. Provinces present almost as great a variety of soil as the hills and table-lands. Each river as it forces a passage through the mountains, falls into the plains charged with its own peculiar sediment, which gives to the soil of the district over which it flows, some distinct property. The deposits of rivers are however, liable to change. Thus, in 1841, a great landslip took place in the gap or defile by which the Indus forces its way through the Himalaya chain, which closed up, for a time, the gap or pass, and converted the upper valley into a mountain lake. The accumulated waters reopened the passage, producing a most destructive inundation, spreading, no doubt, thick beds of transported matter over tracts of country where ordinary floods could not reach. The rise of the Indus at Attock was such as to cause a back current for 70 miles up the Cabool river. The new deposits formed on the occasion would, of course, derive their character from the nature of the rocks involved in the landslip. Thus, no rule can be laid down with regard to the deposits of rivers, or the nature of the various drifted matters of which alluvial beds are made up.

The Jumna, throughout its course in the Doab, is marked by extensive beds of kunkur, or clay charged with peculiar calcareous nodules, totally different from the recent sediment of that river, which consists of fine silicious matter.

The Ganges and the Ramgungah are distinguished throughout Rohilcund by their fine sandy deposit, glistening with mica,—the Goomtee and the Gogra, in Oude, by their fine sandy argillaceous soil,—while the Great and Little Gunduk would appear to have been engaged more recently in the transport of calcareous matter from Nepaul. The Burhampooter and its tributaries, on the other hand, with a larger proportion of iron-sand, contain very little calcareous matter; and from this negative evidence, it was long since inferred, that no great development of limestone rocks was to be expected in the adjacent portions of the Himalaya range, or its prolongation to the eastward of Nepaul; a prediction since confirmed, in a great measure, by the more direct observation of travellers. The great rivers in India which appear at any time to have been engaged in the transport of calcareous drift, are, therefore, the Jumna, and the two Gunduk rivers from Nepaul.

The Delta of Bengal.—The common delta of the Ganges and Burhampooter, comprising Lower Bengal from the sea-face of the Sunderbunds to Morshedabad and Jumalpore, is composed of an admixture of all the sedimentary deposits of the rivers, disposed in beds of clays and sands by the action and reaction of tides and currents, which still flow in all directions through the numerous channels and lagoons which cover a considerable portion of Lower Bengal, the lands being barely raised above the influence of the floods.

We have already passed in review the shallow edges of this great alluvial basin at several remote points, and have thus acquired some knowledge of its structure, where the parts composing it are exposed to light.

The following results of boring operations, penetrating to a depth of 400 feet, throw some light on the various beds of sediment of which the alluvium at its deepest part is composed.

It would be here out of place to enter into details which, however interesting in a scientific point of view, are foreign to the subject on the present occasion. I propose therefore, merely to notice the mineral character of the principal beds of alluvium penetrated in search of water so far as they may appear to bear upon the question of climate and soils.

The surface soil is 10 feet in thickness, followed by adhesive blue clay 40 feet in thickness, containing black peat. This blue clay shows itself in the bed of the Hoogly at low water, 20 miles below Calcutta, where the peat is employed for the purpose of burning bricks. It is this clay which forms the beds of rivers and the basins of lakes and salt-water lagoons, so common throughout the lower portion of the delta.

At a depth of 50 feet from the surface, calcareous and silicious clays, 25 feet thick, occur, containing two layers of calcareous concretionary matter.

Variegated sandy clay occurs, 45 feet in thickness, extending from 75 to 120 feet in depth; at 100 feet from the surface, this clay contains a layer of calcareous nodules.

Loose sandstone and sandy clay, with micaceous slate and nodules of oxide of iron, next follow, to a depth of 150 feet, resting on ferruginous, calcareous, sandy clay, and coarse quartzose conglomerate; micaceous clay, resting on slate-clay, extends to a farther depth of 206 feet.

Ferruginous sandy clays, micaceous clays, with beds of sandstone and loose sand, containing fragments of felspar and granite, resting on shelly calcareous clay, extend to a depth of 382 feet, followed by blue clay containing decayed wood, together with the bones of lizards and chelonian reptiles, teeth of fishes, and bones of a small quadruped.

Lastly, quartz gravel, containing fragments of lignite and coal, blue compact limestone, vesicular trap, and recent wood, occur at a depth of 400 feet, when the work was discontinued.

We thus see, from the character and situation of these beds, that they are a mere attenuation of the same deposits which we find on all sides along the margins of the alluvial basin.

At Benares, Mirzapore, and Allahabad, we find

deep deposits of sand overlying the harder calcareous gravel and clay, forming the more consistent and permanent banks of the Ganges. These are slowly broken down by the force of the stream. Where this occurs, and the waters spread over new lands, we find a nearer approximation to the alluvium of the delta, namely, coarse gravel, alternating with beds of clay and sand. The farther these changes are carried towards the sea, the greater will be the proportion of organic matter mixed up with the sediments.

A great change begins to be perceptible in the surface of the plains, where the river currents are first obstructed by tides. From this point downwards, great deposits of blue clay and mud take the place of sands, and vegetation approaches to the water's edge, forming a striking contrast to the naked beds and plains of sand which are spread far and wide along the course of rivers in the central provinces.

The character and alternation of the more superficial beds composing alluvial plains exercise a considerable influence over soil and climate. It is the ten feet of surface soil, spread over the deep beds of clay, that renders Bengal habitable. Without the surface soil it would be a swamp, and without the clay it would be a desert.

The soil consists of a grey sand, which, becoming saturated with moisture retained by the clay on which it rests, is rendered rich and fertile under cultivation. Neglected, it degenerates into a poor sandy or a heavy clayey soil, which is soon overrun with coarse grasses and other indigenous vegetation, which, when once established, it is difficult afterwards to eradicate.

In Lower Bengal irrigation is not required, and inundations are guarded against by bunds. In the higher parts of the plain, the annual inundations tend to increase the fertility of the lands.

The Danes and French have, perhaps, shown more discernment than the English in the selection of their settlements at Serampore, and Chandernagore, on the right bank of the Hoogly. They have there better foundations and a drier soil to build upon. Although the delta deposits extend several miles to the westward of those settlements, yet the basin becomes rapidly shallower in that direction, and the deposits firmer.

At Ompta, 22 miles west of Calcutta, the tides in the Damooda cease, although they extend up the Hoogly to Kishnaghur, 50 miles north of Calcutta; thus indicating a rise in the surface of the plain of about 12 inches to a mile to the westward, and only six inches in a mile in a northerly direction.

In consequence of the great depth of soft alluvial deposits on which Calcutta stands, heavy masses of buildings have a tendency to sink, which renders the lower apartments in time damp and unhealthy. Rooms on the ground-floor, as well as houses of a single story, have therefore been indiscriminately condemned, although the latter, if sufficiently raised from the ground and built of light materials would, I think, prove far more suitable to the place. Hence the practice of building lofty houses, which obtained for Calcutta its proud title of "the City of Palaces," would seem to be an error. A reference, however, to the neighbouring stations of Barrackpore and Dum Dum, would have been sufficient to show that a lighter and humbler style of building would have been more suitable both to the climate and the soil.

The increased elevation of houses, and the crowding of too many into a contracted space, from the enhanced value of land, tends to increase the evil, which is not easily rectified by drainage; for the settlement of the buildings has a tendency to produce a corresponding sinking of the drains, so that their fall or inclination becomes in time the reverse of what was originally intended.*

Another objection to heavy buildings in Bengal is the attraction of heat from the sun during the day, which the house retains at night, depriving the inmates of the benefit of the cooler temperature of the air after sunset.

These remarks apply equally to barracks and private houses. For the former, brick or iron pillars, enclosed with wooden venetians, surrounded by broad open verandahs and roofed with a thick coat of thatch, would be the most conducive to health and comfort. They should be well raised above the level of the ground, and paved either with Chunar stone or slate —the latter procurable in the Coruckpore hills.

In the eastern parts of Bengal, or wherever the soil is damp or clayey, barracks should be erected on arches or posts raised from the ground.

* For examples of this, I refer to the older houses in and about Hastings Street and Bankshall, to the houses in Park Street and Elysium Row, at the corner of the Circular Road; and, for public buildings, I refer to the Hindoo College, and even to the Government Houses of Calcutta and Barrackpore.

Where hot winds prevail, instead of open venetians, the intervals between the pillars might be enclosed with thin brick walls, having a broad glass door or pair of folding-doors in each space between the pillars, where wet tatties might be employed when necessary —the outer verandah being open.

Besides the objections to heavy buildings above alluded to, their constant pressure on the loose materials of which deep alluvial deposits are made up, produce growing changes beneath the surface, as, for instance, the gradual alteration of a bed of porous materials from a pervious to an impervious state. This would be very likely to be followed by corresponding changes in the soil, such as might seriously affect the salubrity of a place previously healthy, and be attended with increased sickness and mortality. In our ignorance of such matters, we could not be expected to discover the cause, or to trace it to changes beneath the surface: unable otherwise to account for it, it is more than likely we should ascribe the evil to the original defect of the place, and therefore recommend its abandonment, as in the case of Berhampore in 1833, after an original outlay of 3,000,000*l.* sterling; or more recently, the cantonments of Dacca—although both places were, for many years, regarded as healthy stations.

CLIMATE OF BENGAL AND THE N.W. PROVINCES.

CLIMATE depends, as is well known, on solar radiation or the power of the sun's rays, as well as on terrestrial radiation or the evolution of heat from the high temperature of the surface of the land. The first is chiefly felt in the open air, the latter in the shade. Both contribute to produce high temperature, and are influenced in their effects by the presence of marshes, sandy plains, prevailing winds, amount of rain, and other causes of moisture in the soil, and humidity in the air, which it is here proposed to treat separately.

With regard to the power of the sun's rays, they are more fatal to the European constitution in the bright clear atmosphere of marshes in Lower Bengal, than in the more opaque and hazy atmosphere of the N. W. Provinces; although exposure to the direct rays of the sun is dangerous in all parts of India, unless the head is perfectly protected against its effects, particularly over the temples, somewhat on the principle of a native turban.

Terrestrial radiation, which causes that high degree of temperature so trying to European health, unlike the action of the sun's rays, is constant day and night with little variation, its intensity depending on two causes—first, on the soil; and secondly, on the atmosphere—as relative conductors of heat.

It presents the greatest extremes in dry arid or sandy districts. In humid climates on the contrary, while the rays of the sun have less power in raising the temperature of the soil, the atmosphere would seem to have more power in drawing off the increased heat, and thus preventing its accumulation at the surface.

The effect of terrestrial radiation is diminished in proportion as we ascend in the atmosphere; thus, it loses in the Azores one degree of heat for every 145 feet of ascent, and on the Peak of Teneriffe, 190 feet; but these are insular climates, surrounded by sea, which diminishes radiation. In the Alps, 287 feet of altitude is equivalent to one degree of Fahr. In the Andes, the same equivalent nearly has been found to hold good.

In a valuable communication made by Colonel Sykes to the British Association in 1851, it was observed that different months afford different results. Thus Darjeeling, at 7,000 feet, and Sarun, on the plains, differ 17½ miles in latitude; the difference of mean daily temperature in May is 31°·1 Fahr., giving 225 feet for each degree of temperature lost in ascending; but, in December, the difference of mean temperature between the two places is only 14°·8 Fahr., giving 473 feet for each degree of temperature.

Tension, is the term usually, I believe, employed to express, with other properties, that condition of the atmosphere which is found to vary so much in the measure of a degree of heat; but with reference to its influence on climate, we may call this property of the atmosphere, its conducting power with regard to heat.

In comparing Meerut, lat. 29° 1′, with Lohooghat in Kemaon, lat. 29° 22′, the first at an elevation of 1,000 feet above the sea in the plains of the N. W. Provinces, and the second at an elevation of 5,562 feet in the Himalaya, we find that the results afford 255 feet of elevation for every degree of temperature we lose in ascending.

Drawing the same comparison between Darjeeling and Tirhoot, the former situated at an elevation of 7,000 feet, lat. 27° 3′, the latter, in the plains of Bengal, lat. 26° 7′, the height we must ascend for every degree of temperature we lose is 296 feet.

But on comparing Cherraponjee, lat. 25° 16′, at an elevation of 4,500 feet, with Sylhet, lat. 24° 53′, in the plains of Lower Bengal, we must ascend 404 feet (taking one month with another) for every degree of temperature we lose.

It would seem, from the following observations, that the conducting power of the atmosphere, or its power of drawing the accumulated heat from the surface of the earth into the higher regions, depends upon its humidity, as indicated by the fall of rain.

In the instance of Lohooghat and Meerut, where we find the equivalent of one degree of temperature to be represented by 255 feet, the annual fall of rain, taking the mean between the upper and lower station, is 53·16 inches.

In the instance of Darjeeling and Tirhoot, the annual fall of rain, taking the mean between the upper and lower station, is 79·24 inches, the equivalent for a degree of temperature being 296 feet.

Between Cherraponjee and Sylhet, we have an

annual fall of rain, taking the average, in like manner, between the upper and lower station, of 405 inches, with a proportionate increase in the conducting power of the atmosphere, amounting to 404 feet, for one degree of Fahr.

The foregoing results have been drawn from mean annual observations. With regard to the monthly variations, they would appear also to depend—in part, at least—on the changes of the atmosphere in regard to rain. From the following table, the maximum conducting power appears to take place in the month of December, and to diminish towards May or June—holding, however, an intermediate place during the rainy season.

	Kemaon.		Darjeeling.		Cherra-ponjee.		Mean Monthly Oscillations of the Barometer in Calcutta.	Mean Monthly Admissions of Sick for a period of 12 Years.
	Alt. in feet per 1° Fahr.	Fall of Rain, in inches.	Alt. in feet per 1° Fahr.	Fall of Rain, in inches.	Alt. in feet per 1° Fahr.	Fall of Rain, in inches.		
January..	350	6·20	330	3·45	321	0·75	29·937	87·
February.	253	3·71	290	2·40	321	3·05	29·915	105·8
March...	207	1·56	290	4·	401	1·30	29·793	106·5
April....	207	1·17	240	4·45	450	27·60	29·656	116·2
May	228	·95	233	6·75	375	115·15	29·563	127·2
June	207	11·65	370	31·	450	147·50	29·184	124·2
July	304	23·95	310	27·15	409	79·40	29·517	114·2
August ..	267	24·69	310	16·70	450	103·90	29·516	116·2
September	267	5·67	320	19·60	409	71·70	29·662	108·4
October..	267	2·36	285	9·10	450	68·2	29·633	102·3
November	253	·31	370	·10	..	..	29·867	103·2
December	350	·0	402	·10	..	..	29·951	90·2

It is with reference to climate that this question assumes so much importance; for, although temperature at the surface may be but little affected, yet it is easy to see how the condition of the air may become affected, more especially as regards contagion, by its conducting powers. It is worthy of remark, that, during the hot months, when cholera prevails, as well as small-pox, and jungle fever is the sure consequence of visiting certain districts, the conducting power of the atmosphere is at its lowest point. Nor are we to overlook the solution here offered to the common remark, that a fall of rain clears the air, and is supposed to check the spread of cholera.

From registers of the barometer kept at Delhi, Mozufferpore, Assam, and Calcutta, we find the mean monthly pressure at each of these places to present the highest standard in December and January. The lowest in Calcutta is in June, at Delhi and Assam in July, and in Tirhoot in August.

In the last column of the foregoing table will be found the monthly admissions of sick into hospital, on an average of twelve years, from 1,000 Europeans troops stationed in Calcutta. The scale of sickness is thus shown to follow, in an inverse ratio, the conducting power and pressure of the atmosphere.

The admissions of fevers show this is in a still more forcible light. The increase from 231 for January, to 488 for June is steady and progressive during the intermediate months, and from June we have a progressive diminution to January.

There are facts with regard to fever which prove it to be independent of temperature alone.

Thus, a party, consisting of four Europeans and about fifty natives, were seized with fever in the

Rajmahal Hills between the 11th and 14th of March. On the other hand, the Goorkah force, consisting of ten thousand men, crossed the Tarai on their return to Nepaul as late as the 6th of June, very few cases of fever having occurred.

In the Rajmahal Hills, fever sets in when the air is thought dry, as well as the stiff hard laterite soil; although the latter is in reality still charged with moisture from the preceding rains, and which it parts with slowly after it has become heated to a certain point about the middle of March: at this time, such tracts become unhealthy.

In other districts, composed of lighter sedimentary soil, such as the Tarai and the Sunderbunds, fever does not set in until after the first fall of rain, when the malaria has immediate effect. In both cases, it loses its effects for a time when the rains have fairly set in, reappearing when the fall of rain becomes interrupted towards the close of the rainy season, when the sky becomes bright and clear. The districts composed of laterite or heavy clay, take a longer time to dry after the rains have subsided, and are not safe to enter until the middle of January. The districts composed of sedimentary deposits, on the other hand, become safe after November.

Winds.—The north-east monsoon begins about the middle of October, and continues until the end of February or middle of March, corresponding with the cold season in Bengal. It is a cold dry wind, which prevails over the whole of Central India and the N. W. Provinces, accompanied with a bright clear sky. Nothing can exceed the bracing effects of this season on the general health and spirits of European residents in India. To those who are subject to organic disease of any kind, it is, however, very trying; and the bills of mortality show a higher ratio at this, than any other season, more especially among those who suffer from organic affections. The reason of this, no doubt, will be found in the great diurnal variation in the range of the thermometer, amounting to as much as 30 degrees between the temperature of the day and night. These diurnal extremes are greater in the dry open plains and table-lands than in the eastern districts of Bengal, where, although the mean annual temperature is lower, yet the mean monthly temperature is higher during the cold season. Thus, the districts of Purnea, Tirhoot, and Goruckpore, which present a mean annual temperature of 2°·4 less than Allahabad, Cawnpore, and Delhi, show a mean temperature, for the month of January, 4° Fahr. higher. The extremes of heat and cold, are consequently less in the eastern districts; which may be accounted for by the fact of these districts being sheltered by mountains from the north-east monsoon.

Thus, at Mozufferpore in the months of December, January, and February, we have 61 days of west, and 28 days of mild east wind, and no northerly wind whatever during a period when cold north-east winds prevail throughout other parts of the Gangetic plains and Central India. The same observation applies to Assam, Sylhet, and Purnea, as well as the eastern districts, which thus present a milder climate than

Central India and the plains along the course of the Ganges.

The close of the north-east monsoon in Lower Bengal is usually marked by dense mists occasioned by the south-west wind, which sets in as a sea breeze from the south about the end of February. This mild humid wind coming in contact with the cold north-east wind, a sudden condensation takes place early of a morning in the form of mist, which lasts until dispelled by the sun's rays about eight or nine o'clock. The same effect is sometimes produced earlier in the season, usually about the end of November or beginning of December, occasioned by oscillations between land and sea breezes. These changes are marked by prevalence of colds.

The hot winds set in about the beginning of March, and continue for three months, with increased intensity, until the first heavy fall of rain that takes place in June. During their continuance, the surface of the country becomes dry and parched; the tanks and streams dry up, and water becomes scarce and often impure; vegetation also becomes in some degree suspended, with an almost total absence of green pasture. Even the atmosphere loses its transparency, and becomes thick and hazy from the quantity of impalpable earthy matter floating in the air. It is true, there are a few hours of a morning, up to eight or nine o'clock, to attend to out-door business daily, before the wind sets in and the sun becomes too powerful. But the rest of the day must be spent under cover, and, if possible, in a house, well-secured against the admission of air, unless it is made to pass through one of the windows enclosed with wet grass instead of glass; which, with the punka, serves to keep the temperature of the house five or six degrees below that of the external air. After sunset, when the wind subsides, the doors may be opened to air the house. The air itself of an evening is close and sultry, and everything exposed to the hot parching sun and wind during the day—such, for instance, as houses, trees, and the like —emit a glow of heat, so that it is not until the morning that things have cooled down sufficiently to allow of exercise in the open air.

Except the prostrating effects of the hot winds on European energy, the season is not considered unhealthy; unless, indeed, when cholera appears. The hot winds extend over the whole of the central provinces, from Orissa to the Punjab, and are severely felt along the plains of the Ganges.*

There are, however, certain provinces, namely, the eastern parts of Rohilcund, Oude, Bahar, and Eastern Bengal, where the influence of these winds may be said to cease. The prevailing wind at Mozufferpore is from the east. It blows strong from the west in February and March. North winds are of rare occurrence.†

The hot west wind is, indeed, nothing more than the south-west monsoon drawn from its natural course by the radiation of heat from the hot and parched plains.

The inhabitants of Assam do not feel it, because, from other local causes, the wind blows steadily down

* The three months of April, May, and June, present a higher mean temperature at Balasore, in Bengal, by 2°·9 Fahr., than at Ferozpore, in the N. W. Provinces.

† Dashwood: *Journ. Asiat. Soc.;* Bengal, vol. iii., p. 79.

the valley of Assam from the elevated regions lying to the south-east. They therefore enjoy, during the months of April, May, and June, a comparatively cool refreshing breeze, which secures to them a mean temperature, during the period of the year in question 15°·6 Fahr. cooler than Allahabad, Cawnpore, and Delhi, in the same parallel of latitude. In like manner, the south-west monsoon reaches the eastern parts of Bengal fresh from the sea, securing to the inhabitants of Chittagong, Dacca, and Sylhet, a lower temperature, by 9° Fahr., than that of the western districts of Midnapore, Burdwan, and Beerbhoom, during the months of May and June. Here, receiving fresh impetus from the Assam current, it passes over the districts of Tirhoot, Champarun, Goruckpore, and the eastern parts of Oude, making its influence felt as a humid south-east wind. This wind is followed in its course by the grasses and other vegetation of Lower Bengal, indicated by the green verdure of the forests, and the presence of the khajur, or date palm, the sissoos, and other trees of damp soil and humid atmosphere, which are found to characterise the forests to the 27° N. lat., or five degrees beyond the range to which these features extend on the western side of Bengal.

The traveller from Calcutta to the Upper Provinces loses at Burdwan the vegetation of Lower Bengal, and fancies he has left it behind him, until he is led, perhaps, by accident, from Lucknow to Fyzabad, when, within the short distance of 70 or 80 miles, the landscape suddenly changes, and from parched plains he enters a climate of perpetual spring.

The following entries from a note-book will make this more clear, perhaps, than lengthened descriptions:—

"*April 22nd.*—Proceeded from Nawabgunge to Subdalgunge; distance, 8 miles. The aspect of the country is quite changed since leaving Lucknow; the groves, hitherto composed of mango trees, are now interspersed with sissoos and a dense green underwood."

"*May 1st, Fyzabad.*—Tamarind trees here take the place of mangoes, in the formation of topes; and the date palm, not seen to the westward in districts exposed to hot winds, as well as other trees of similar habit, here make their appearance."

Subsequent experience, during the month of May and beginning of June, served to confirm the above observations in proceeding through the districts east of the Gogra.

The foregoing remarks agree generally with observations made at Mozufferpore, where the winds are stated to be easterly throughout the months of April, May, and June, at a time when hot westerly winds prevail in other parts, and with the tables kept at Kulsea, at the head of the Doab Canal, where easterly winds to westerly are as 3 to 1.*

Temperature and Rain.—We shall find, on going into this important subject, that the tracks of the prevailing winds in the plains of Bengal and the N. W. Provinces, present, in these plains, two separate ranges of climate, distinguished from each other by a difference of mean annual temperature amounting to no less than 5°·4 Fahr., and a difference of no less than 51·19 inches in the annual fall of rain.

* R. B. Smith, Bengal Engineers: *Calcutta Journ. Nat. Hist.* v. 413.

These two climates have never been clearly pointed out, nor taken advantage of in the selection of stations for European troops.

Umballa and Ferozpore, in the same parallel of latitude, present a difference of 1°·6 Fahr. in favour of the former.

	Ferozpore.	Umballa.
January	55·9°	57·°
February	62·5	60·3
March	72·3	73·2
April	84·	81·2
May	92·8	87·9
June	97·5	96·3
July	88·3	87·1
August	93·3	90·7
September	94·	91·6
October	86·	84·3
November	70·8	67·1
December	61·	62·2
Mean annual temperature	79·8	78·2
Mean annual fall of rain	23·13	25·76

The coolest month at Umballa is 57°; the hottest 96°·3.

The coolest month at Ferozpore is 55°·9; the hottest 97°·5.

In considering the question of temperature, with reference to the European constitution, it is not the mean temperature of the year, so much as that of the hot season, and close of the rains, that we have to consider.

Now, we find that the mean temperature of Umballa is three degrees less than that of Ferozpore during the hot months of April, May, and June, and that there are 2·63 inches of rain more at Umballa than at Ferozpore.

The annexed columns exhibit the results of the registers of three stations situated on the western side of the Gangetic plains, and corresponding in their range of climate with Ferozpore, reduced to their mean monthly and mean annual temperatures, and compared with the results from three stations situated on the eastern sides of these plains, corresponding in their range of climate with Umballa.

	Delhi, Cawnpore, and Allahabad.	Seharanpore, Moradabad, and Bareilly.
January	60·3°	56·1°
February	66·2	60·3
March	76·2	70·
April	89·3	78·6 }
May	98·0	87·1 }
June	96·	93·1 }
July	85·3	85·8
August	89·6	88·3
September	85·8	82·
October	80·	79·8
November	72·1	66·7
December	64·3	59·1
Mean annual temperature	80·5	75·2
Mean annual fall of rain	28·46	43·85

We thus find the mean annual temperature of the three stations of Seharanpore, Moradabad, and Bareilly, situated on the north-eastern border of the plains, to be 5°·3 Fahr. less than that of Delhi, Cawnpore, and Allahabad, and 7°·9 Fahr. in favour of the first-mentioned stations during the hot months of April, May, and June.

We also find that the annual fall of rain, which contributes to the amelioration of great heat, is 15·39 inches more on the north-eastern side of the plains, than on the north-western.

In the drier climate of the western stations, we have 10° Fahr. between the mean temperature of February and March, 13°·1 between March and April, and 9°·3 between April and May, which is the hottest month. In the more humid climate of north-eastern stations, the accession of heat is more gradual, and does not reach its maximum till June; while the mean temperature of the three hottest months is 8°·2 Fahr. cooler in the north-eastern stations: and the months of April and May nearly 11° cooler than the corresponding months at Delhi, Cawnpore, and Allahabad.

We still perceive, in comparing the annexed columns, the same decided balance in favour of the eastern districts, in the lower mean annual, as well as in the mean monthly temperature.

	Ghazipore, Dinapore, Bhagulpore.	Goruckpore, Mozufferpore, Purnea.
January	64°·9	62°·1
February	69·4	65·3
March	79·1	70·7
April	87·4	85·
May	94·6	89·
June	91·8	87·6
July	87·	85·7
August	86·	85·6
September	85·4	85·
October	82·3	80·2
November	70·6	70·3
December	61·1	63·
Mean annual temperature	80·1	77·6
Mean annual fall of rain	35·04	49·99

We also perceive it in the more gradual approach of the hot weather, there being in the Gangetic stations a difference of 9°·7 Fahr. between the temperature of February and March, 8°·3 between March and April, and 7°·2 between the mean temperature of April and May; while in the north-eastern districts the corresponding monthly increase is 5°·4, 4°·3, and 4° Fahr. from February to May.

In the Gangetic stations there is a difference of 33°·5 Fahr. between the mean maximum in May and mean minimum in December; while in the north-eastern districts there is only 26°·9 between the mean maximum and mean minimum.

If all the different stations were compared singly, the results would be the same, because there is a general correspondence throughout the climate of each group. Thus, the mean annual temperature of the three stations of Goruckpore, Mozufferpore, and Purnea, is 78°, 77°·8, and 77°·7 Fahr. respectively, showing only a difference of three-tenths of a degree between them. In like manner, the mean annual temperature of Ghazipore, Dinapore, and Bhagulpore, is 80°·4, 79°·8, and 80°·5, showing only seven-tenths of a degree of difference.

The difference of temperature already shown to exist between the eastern and western plains, becomes more strongly marked in Lower Bengal, where the eastern plains will be found, by a comparison of the annexed tables, to present a lower mean annual temperature, by 3°·4 Fahr., than those which lie on the western side, although both are within the same parallel of latitude. The plains of Mymensing, Sylhet, and Dacca, also present the important advantage of a lower mean monthly temperature, by 5° Fahr., than the plains of Burdwan and the stations to the westward, during the five hot and trying months of May, June, July, August, and September.

	Beerbhoom, Burdwan, Bancorah, Midnapore.	Mymensing, Sylhet, Dacca, Chittagong.
January	70°·1	61°·1
February	73·1	69·5
March	80·5	78·6
April	80·7	80·2
May	91·5	83·3
June	88·5	81·
July	84·6	82·8
August	86·1	83·
September	86·4	83·7
October	81·	79·8
November	72·2	74·1
December	69·1	68·1
Mean annual temperature	80·5	77·1
Mean annual fall of rain	28·52	118·20

The mean annual temperature of the stations in the western group is as follows:—Beerbhoom, 81°·3; Burdwan, 81°·1; Bancorah, 82°·9; and Midnapore, 76°·9 Fahr. That of the eastern group differs little in the different stations. The mean annual temperature of Mymensing is 76°·9, that of Chittagong 76°·2, Sylhet is 77°·8, and Dacca 79°·7 Fahr. The mean annual fall of rain in the eastern district is 118·20 inches; in the western, 28·52 inches.

The valley of Assam corresponds in climate closely with the eastern group of stations of Lower Bengal. The mean annual temperature of the valley, deduced from that of the stations of Gowahatti, Goalpara, Teezpore, Seebsagur, and Cachar, is 76°·2 Fahr.; and the mean annual fall of rain 84·11 inches, although it differs much in certain localities. Thus, at Goalpara, the annual fall of rain is 116 inches; at Cachar, on the opposite side of the river from Goalpara, it is 102·84 inches; at Gowahatti, it is 52·74 inches; Seebsagur, 85·18; and Teezpore, 63·49.

GENERAL RESULTS.

The mean annual temperature of the western plains from Ferozpore through Delhi, Cawnpore, Allahabad, Ghazipore, Dinapore, Bhagulpore, and along the right bank of the Hoogly to Midnapore, is 80°·2 Fahr. That of the eastern plains, from Umballa through Seharanpore, Moradabad, Bareilly, Goruckpore, and Tirhoot, to Mymensing and Dacca, is 75°·5 Fahr.

The mean annual fall of rain in the western plains is 28·78 inches; that of the eastern plains, 59·32 inches.

The prevailing winds in the western plains are northerly from October to the end of February, when hot westerly winds set in and continue until June, they then become variable until October.

The prevailing winds in the eastern plains are easterly from March until October, when westerly winds prevail from October until the end of February.

The boundary between the eastern and western plains thus characterised by climate, may be generally defined by the Hoogly river to Morshedabad, from thence by the Ganges to Dinapore, and from thence by a line extended to Meerut.

The mean annual temperature of Assam is 76°·2. The hottest months are July, August, and September, when the mean temperature is 83°. The months of April, May, and June present a mean temperature of 79°·2.

Table-lands of Central India.—With regard to the relative temperature of the stations situated on the heights as compared with the plains, the annexed table shows a difference of 3°·9 between the mean annual temperature of Hazareebagh, at an elevation of 1,500 feet above the sea, and Morshedabad, situated in the plains, at the foot of the table-land, with a difference of only 11 miles of latitude between them.

	Hazareebagh.	Morshedabad.
January	64·3°	63·°
February	65·4	65·5
March	76·8	79·2
April	83·	83·
May	89·1	88·
June	86·2	85·5
July	79·9	82·7
August	79·3	88·
September	78·8	88·
October	73·3	83·
November	69·5	75·
December	64·7	68·
Mean annual temperature	75·8	79·7
Mean annual fall of rain	32·06	40·48

The hot weather sets in uniformly at both places, and there is little difference until July, after which there is a mean monthly difference of 10°·5 Fahr. in favour of the upper station during the ensuing three months. It is cooler by one degree than Goruckpore and the eastern plains, during the hot season, and four degrees cooler than Dinapore, Ghazipore, and Bhagulpore, during the hot weather. It is, however, ten degrees cooler than any station in the plains from August to October, which is sufficient in itself to stamp the superiority of Hazareebagh over every station in the plains, for the cantonment of European troops.

The mean temperature of Hazareebagh, Saugur, and Baitool, three stations on the table-lands, during the months of August, September, and October, is 78°·3, while that of Dinapore, Morshedabad, and Midnapore, for the same months, is 84°, showing a difference in favour of the table-lands in general, of 5°·7. Again, the mean temperature of the following four stations on the table-land,—namely, Saugur, Baitool, Neemuch, and Hazareebagh—during the hot months of April, May, and June, is 86°·8 Fahr., while that of Dinapore, Morshedabad, and Midnapore (taken collectively), situated on the plains under the same general parallel, is 87°·8, or only one degree higher than that of the table-lands. It is, therefore, at the close of the rainy season that the table-lands present the advantage over the plains, and those advantages are most marked in the instance of Hazareebagh.

The general temperature of the table-land is distinguished from that of the plains by the shorter duration of the hot weather, and the lower temperature of the nights—advantages ill-expressed by these tables of the thermometer, which were only noted in the day time.

The north-western portion of this platform lying between the Chumbul river and Ajmere, owing to the more scanty fall of rain, its lower levels, and light sandy soil, presents a higher temperature than either Rajpootana or Central Malwa. In these last tracts, as well as throughout the southern division of the Saugur and Nerbudda territories, embracing the stations of Mhow, Saugur, and Jubbulpore, the soil is more retentive and clayey, and the climate is characterised by a more copious fall of rain—advantages which have their drawbacks in the greater prevalence of fevers in these parts as compared with the northern districts about Gwalior and Ajmere.

The prevailing winds are westerly, varying, in the rainy season, a few points to the south, and, during the hot season, to the north. In the cold season, the winds are usually north, changing occasionally to the south-east.

These observations apply only to the lower elevations of Central India, the higher parts not being occupied.

At Gurgaon, near Delhi, according to the observation of Major Oliver, May is the driest month, August the dampest. But, considering the comparative temperature, there is a less quantity of aqueous vapour in a given quantity of air in January than in May. He found, by a rough calculation, that the weight of aqueous vapour in a cubic foot of air varies from 3·3 grains in January to 10·3 in August.

Comparing the driest month at Delhi with the driest in Calcutta, he found the ratio to be as 5 to 3, that of the dampest, as 5 to 4.

Meteorological Register, kept at Calcutta, from the 1*st November*, 1843, *to* 31*st October*, 1844.

Such registers are intended to afford materials for the investigation of climate; and if held available, in the places where they are kept, to persons engaged in researches of this nature, publication may be well dispensed with. They are bulky if given in detail, and useless if too much abridged.

The prevalence of small-pox and cholera in March and April, and of a peculiar form of brain fever which set in in the early part of the rains, may render the subject more interesting as accounting for those complaints by the unusually early setting in of the hot season, its violence and protracted length. The late and irregular appearance of the rains, with the heavy falls of rain in August and September, and the consequent diminished temperature of those months, will account no less satisfactorily for the unusual healthiness of that part of the season.

The observations were made at sunrise, 9h. 40m., noon, 2h. 40m., 4h., and at sunset; in all, at six different periods. The lowest daily temperature is at sunrise, the highest at 2h. 40m., and this holds good throughout the year. The lowest temperature observed was on the 19th January, when the thermometer stood at 51°·7; the highest was on the 10th of April, when the thermometer stood in the shade at 104°. The mean temperature of December, the coolest month of the year, was 72°·02; and of April, the hottest month of the year, 89°·6; the mean temperature of the whole year, founded on six daily observations at the hours above stated, is 82°·35. The mean minimum is 81°·15, and the mean maximum is 93°·67.

The greatest extremes were in the months of February and March, when there was a difference of 38° between the maximum and minimum temperature observed. The months of August and October afford the nearest approximation to the mean temperature of the whole year.

The following Tables exhibit the thermometric and barometric results for the several months:—

Thermometric Results.

Months.	Minimum.	Maximum.	Month's variation from the annual mean.	Epochs. Of min.	Epochs. Of max.	Difference of extremes, monthly.
November	60°	86°2	— 3·91	29th	6th	26·2
December	53·	84·5	—10·37	17th	4th	34·1
January	51·7	84·2	—10·01	19th	25th	32·5
February	56·	94·	— 4·32	1st	26th	38·
March	63·	101·	+ 4·44	1st	28th	38·
April	71·9	104·	+ 7·21	29th	10th	32·1
May	76·	98·6	+ 5·34	14th	6th	22·6
June	76·	95·	+ 4·46	13th	20th	19·
July	77·	93·2	+ 1·61	1st	24th	16·2
August	76·9	90·5	+ 1·31	14th	25th	13·6
September	77·	96·	+ 3·16	9th	27th	19·
October	73·	94·	+ 1·26	31st	6th	21·
Means	81·15	93·67				31·23

Barometric Results.

Months.	Extreme minimum.	Extreme maximum.	Monthly difference from the annual mean.	Epochs. Of min.	Epochs. Of max.
November	29°785	30°042	+·160	17th	2nd
December	29·830	30·094	+·244	1st	19th
January	29·772	30·152	+·230	26th	17th
February	29·770	30·062	+·208	27th	5th
March	29·606	29·978	+·086	29th	1st
April	29·458	29·834	—·051	2nd	16th
May	29·385	29·734	—·144	29th	1st
June	29·271	29·694	—·220	29th	18th
July	29·326	29·713	—·190	28th	21st
August	29·196	29·740	—·191	21st	30th
September	29·394	29·865	—·045	8th	29th
October	29·572	29·937	—·074	12th	28th
Means	29·530	29·903			

The mean height of the barometer during the year was 29°·707. The lowest mean monthly height was 29°·484 in June. The highest mean monthly is 29°·951 in December; the lowest extreme, 29°·196, was on the 21st August, and the highest, 30°·153, on the 17th of January. September affords the nearest approximation to the mean annual. The maximum pressure occurs about 10 A.M.; the minimum at 4 P.M., except on the 26th of January, 27th of February, and 29th of May, when the minimum pressure was observed at 6 P.M.

Rain fell on 101 days during the year to the amount of 74·72 inches, of which we had 65 inches, or $\frac{5}{6}$ths of the whole, in 72 days of June, July, August, and September, and the remaining $9\frac{3}{4}$ inches only throughout the rest of the year. To this peculiarity, combined with the diminished fall of rain during the preceding year, which only amounted to 63·34 inches, may probably be ascribed the excessive temperature of the months of March and April in particular, and the unhealthiness that prevailed at that period in consequence.

The following Table exhibits the peculiarity of wind and weather generally throughout the year:—

Quantity of Rain and variations of Wind.

Months.	Weather.				Of the Wind at Noon.									
	Quantity of Rain.	Number of Days on which Rain fell.	Number of Days on which no Rain fell.	Days on which it Thundered.	N.	N.E.	N.W.	S.	S.E.	S.W.	W.	E.	Calm.	No. of Days of Observations.
November ..	0·	0	29	0	16	3	7	0	0	1	3	0	0	29
December ..	0·86	2	29	0	15	4	6	0	0	0	3	1	1	30
January....	0·22	1	30	0	15	2	4	3	0	0	6	1	0	31
February ..	0·08	1	28	0	19	3	4	4	0	2	6	1	0	29
March	0·22	1	30	0	0	2	1	8	0	10	10	0	0	31
April......	3·13	6	24	1	0	1	0	17	1	3	6	2	0	30
May	7·44	12	19	3	0	3	1	20	1	4	0	2	0	31
June	12·13	14	16	3	1	2	2	11	1	7	3	3	0	30
July	13·72	23	8	1	0	1	1	10	6	3	4	5	0	30
August	26·61	23	8	4	1	0	1	11	6	4	1	7	0	31
September ..	5·02	12	18	3	0	2	4	7	2	6	6	3	0	30
October ...	4·99	6	14	0	1	3	3	0	2	2	5	4	0	20
Totals ..	74·72	101	252	15	68	26	34	91	19	42	53	29	1	353

The following Table exhibits the days and hours on which it thundered during the year:—

Periods at which Thunder occurred.

Hours.	November.	December.	January.	February.	March.	April.	May.	June.	July.	August.	September.	October.
Sunrise..	..	..	..	..	..	..	..	..	..	..	..	..
9 h. 50 m.	..	..	..	..	..	..	..	..	..	..	..	..
Noon ...	..	..	..	..	..	30th	18th 19th	..	..	18th 19th	..	..
2 h. 40 m.	..	..		..	..	..	18th 30th	24th	3rd	6th 11th	4th 12th	..
4 h.	..	..	..	..	..	..	..	..	..	..	..	..
Sunset ..	..	..		..	..	..	..	2nd 7th	..	..	7th	..

Thunder, it will be observed from the above Table, occurred on 15 days from April to September, while the rest of the year was exempt from it; on five of these occasions it occurred at noon; on seven it occurred at at 2h. 40m.; and on three occasions at sunset.

VEGETATION AS AN INDICATION OF CLIMATE.

The forests composing the Sunderbunds and borders of the coast from Bengal to Pegu, consist of Sonerilla, Heretiera, Rhizophora, Soneratia, and thorny palms, as Phœnix paludosa, Calami or ratans, and other species which never travel far from the coast. These are mixed with Acacia, Lagerstræmia, Barringtonia, Dillenia, Dalbergia, and Nauclea,—which extend further inland, with cocoa-nuts and date-palms, marking the sites of villages.

This vegetation passes into the interior, following damp and humid tracts, but terminating abruptly wherever the influence of hot dry land wind extends, or where the level of the land rises much above that of the sea.

The traveller from Calcutta thus loses Dalbergia sissoo, the cocoa-nut, and date-palm, as well as Heretiera literalis and other species common about Calcutta, before he proceeds much farther than Burdwan, and from thence to Lucknow, he sees them no more.

The forests on the first part of his way are composed of some two or three kinds of Boswellia and Diospyros, as well as of Conocarpus and Pentaptera, with Odina wodier; and as the country opens, he finds it wooded with both Bassia latifolia and Terminalia bellerica, which, from their oak-like appearance, give a domestic character to the wildest tracts. These features alternate with others formed by Butea frondosa, on which the lac insect feeds, and which usually takes entire possession of the tracts it occupies, almost to the exclusion of other prominent plants and trees, forming of itself a wild straggling brushwood.

The peepul tree Ficus religiosa, occasionally presents a fine object; and the eye is charmed, on the first part of the way, with the bright fuchsia-like flowers of Grislea tomentosa. Even that he loses long before he crosses the Sone river. But fine peepul and mango trees still continue, and the neighbourhood of towns and villages may be discovered at a distance by the tall fan-leafed palm, Borassus flabelliformis, which abounds in their vicinity.

What vegetation wants in variety is made up, as you get into Oude, by the grouping of the mango trees into topes or square formal plantations, in which the trees are grown in close lines, affording a cool shade from the hottest sun; there is little green sward, and the ground below is entirely free from the obtrusion of underwood.

In this way, the deficiency of vegetation in the upper provinces, but especially in Oude, has been provided against. Still, notwithstanding all that art can do, the country presents, on all sides, a dreary and parched appearance, especially in the hot weather.

Proceeding from Lucknow to Fyzabad, a change is observed. Dalbergia, Lagerstræmia, Caralia lucida, Bignonia, and Ficus indica, begin to appear in mango topes, with an underwood composed of Justicia adhatoda, Mimosa scandens, and other plants of Lower Bengal. The change is rendered still more remarkable by the re-appearance of the date-palm, Phœnix sylvestris.

At length it becomes clear that, in scarcely 50 miles, we have changed our climate; and this fact is further brought home to our conviction by the busy hum and joyous notes of thousands of birds, which congregate in the districts east of the Gogra river, to avoid the hot winds.

Such marked and sudden limits in the distribution of species thus brought together, has not been unobserved by many distinguished naturalists, who have usually ascribed it to the effect of longitude, which in Central Asia was supposed, from some cause, to exercise a more powerful influence than elsewhere. The real cause will however, I think, be found partly in the local peculiarities of climate above explained.

The agricultural produce of Lower Bengal is chiefly rice, together with coarse fibre-yielding plants, affording *sun* or Bengal hemp, and another coarse hemp-plant, which yields the material from which gunny-bags are made.

A little higher, or about the parallel of Calcutta, the cultivation of indigo begins in the Sylhet and Burdwan and intermediate districts. From thence upwards, sugar-cane begins to be cultivated as well as indigo, gradually extending into Assam. *Urrhur*, or *dhall*, a kind of pulse, begins also to mingle with the rice cultivation as we proceed upwards. Flax, as an oil-seed, also makes its appearance, grown in drills with the field pea. As we advance towards Goruckpore, other pulses become general crops in the cold season, as well as wheat, which is sown in October, after the rains have subsided, and ripens in February. The poppy is also extensively cultivated in Goruckpore and the neighbouring districts, where the Benares opium is chiefly produced. The Patna opium is chiefly the produce of the Bhagulpore and Monghyr districts, on both sides of the Ganges.

The tea plant of Assam is now extensively cultivated in a light grey soil, resting on a fine sandy subsoil. The same plant has been recently found wild in the valley of Cachar, lying to the east of Sylhet, where extensive arrangements have been made for its cultivation, of the ultimate success of which there cannot be the slightest doubt.

Low hills of granite, capped with sandstone, are seen rising abruptly from the plains, 12 miles west of Banda. Here cotton is cultivated. The roads are dry; and rocky elevations give a charm to the scene, after having been accustomed to a long residence in the plains. A few miles after passing Adjeegur hill and the hill of Callinger, you come to the pass, formed of trap, leading to the upper districts of Bundelcund.

The prevailing soil, all the way from Allahabad to the pass, as well as over it into Malwa, is black and coarse, consisting of a large proportion of sand and clay. It retains moisture more than the light sandy soil of the plains, and becomes soft and muddy in the rains. Its fertility is proverbial; "but, as might be expected," says Dr. J. Adam, "agues are very common in the whole of the low district of Bundelcund, and sometimes so severe in European habits as to require a change of air for their removal." The natives, however, are not more subject to ague here than in other parts.

It is in this stiff black soil that the *Purwa* cotton is grown in Bundelcund. The *Narma* cotton of Guzerat grows in any soil, but is said to prefer light, to the heavy black soil.

In the table-lands, cultivation is more limited than in the plains. Rice is only cultivated in small valleys, where means of irrigation are available. Wheat is grown in the dry season as in the plains, together with Phaseolus "koortee," Lathyrus, Cicer orientalis, and Cytisus catejan, which yield various kinds of pulse. The oil-seeds grown in the table-lands, as well as in the plains are, besides mustard and flax seed, Ricinis communis or castor oil seed, helianthus, an oil-seed — called in the hills, *Sarrha Soojah,* and Sesamum muralis or *teil.*

Sorghum vulgare is almost the only object of cultivation by the Santals of the higher Rajmahal hills, where it is sown in the deepest forests during the rains, with very little preparation of the soil, other than merely removing so much of the forest as obstructs the rays of the sun. The ground, which receives very little attention, does not yield a second crop. The grain is preserved in the ear until it is required for use. It is the coarsest description of grain, but is yielded with the least labour or exertion, and is, therefore, suited to the simple habits of the aboriginal hill tribes, who hold themselves as distinct as possible from all foreign intercourse, as well as from their neighbours in the plains.

In Pegu, vegetation presents the same leading features as in Bengal, with fewer breaks or interruptions to the more equal diffusion of species, arising from variation of climate. In the delta of the Irawaddi we observe the same net-work of creeks, separated by low forests, subject to the influence of tides. The vegetation is generally the same as we observed in the Sunderbunds of Bengal, with a larger diffusion of Rhizophora, screw pines, and ratans. Of the latter, Zalacca gives a new character to the native dwellings and houses of all kinds in the eastern provinces by its leaf—universally employed in thatching.

At Rangoon, we find many ornamental plants, which in Calcutta can only be preserved with the greatest care as small shrubs, assume the character of middle-sized or large trees, as several species of Calophyllum, Michelia, and Mesùa férrea. The palms, more numerous than in Bengal, consist of Corypha taliera and elata, the sago palm (Saguerus Rumphii), Licualis peltata, and spinosa, besides the cultivated Areca catechu, Borassus, and cocoa-nut; these, together with the lofty wood-oil trees (Dipterocarpus), towering with their tall naked trunks and bushy crown, high above the forest, give a peculiar character to the scene about Rangoon, which is much heightened by the numerous pagodas. The character of the forests, as they extend up the valleys of the Sitang and Irawaddi to the north, assimilate with those of Assam and other parts of Bengal. The agricultural capabilities of Pegu are also very great, and only limited by the want of sufficient population and the consequent high price of labour.

Rice, timber, and catechu, are the chief produce. Rice cultivation is that which is most likely to extend, the soil and habits of the people being formed and adapted to it in Pegu.

In the hilly tracts between the Sitang and Irawaddi, the soil and climate are both well suited to the growth of cotton, as the former presents all the varieties of the Malwa and Guzerat cotton soils, while the climate, upon the whole, is better than that of the cotton districts of Bundelcund. Indeed, I think some very favourable climates might be found on the northern frontier of Pegu, in the Yomah hills, between Prome and Tounghoo. Cotton is already grown here to a small extent, chiefly for local consumption. The want of population, and the consequent high price of labour in Pegu, will, however, always operate to its disadvantage, as compared with Bengal, more particularly as regards such articles as require much labour or skill in their production, as sugar, and indigo. Some of the native plants yielding fibre might succeed in Pegu, where little labour would be required in their cultivation.

GOITRE AND CRETINISM IN KEMAON.

THE following observations on the subject of goitre and cretinism, which were printed in India, some years since, may give an importance that was wanting to the geological descriptions contained in the first part of these pages, and be the means of drawing the attention of medical topographers more closely to the influence of soils on the human constitution:—

When engaged in collecting information given in the preceding pages, I was struck with the frequency of goitre in one portion of the district; while the other was almost perfectly exempt from the complaint. The Alpine characters of the province are the same in every part, the inhabitants all belong to the same tribes of Hindoos, and are subject to fewer irregularities in their mode of life than any other people in the world. In such a field, there could be little merit in eliciting highly important facts connected with this intricate subject.

That portion of Kemaon which lies on the south side of the Ramessa river, is composed of silicious and argillaceous rocks. Granite, which penetrates through the upper beds, forms a lofty ridge about eight thousand feet high. In the centre of this ridge there are numerous small valleys, some of them seven thousand, and others as low as three thousand, feet above the sea. These are inhabited by persons who, to procure pasture for their cattle in winter, descend into the plains, and are absent from their villages for five months every year. From inquiries made amongst these people, I found them to be affected with goitre in the proportion of one in five hundred; but as they do not constantly reside in the mountains, they are excluded from the more minute statistical details on the subject.

On the north-eastern side of the chain of mountains above mentioned, are numerous deep river-valleys and ravines, as well as low mountain ridges, which afford a climate more congenial to the feelings and wants of the inhabitants, who here reside constantly in their villages. Of these villages, forty-six have been visited; but two of their number, having been only occupied for three or four years, are not included; so that the number of villages on the south of the Ramessa river, which we are now to consider, amount to forty-three, and contain a population of 3,700: of this number I found only seventeen persons affected with goitre, and these were exclusively adults. The localities of these villages are as diversified as can well be imagined. Some are erected on narrow ridges, others in deep valleys, surrounded by abrupt and lofty mountains; others on rugged declivities, between lofty peaks on the one side, and dark ravines on the other, into some of which the sun but partially penetrates. The different altitudes of these villages vary from two thousand to six thousand feet.

Let us now cross the Ramessa river, and enter the district of Shore, whose geological distinctions have been pointed out in a former part of this work; and we shall find that an eighth part of the people are affected with goitre. Yet all the inhabitants of the province are of the same religion: they intermarry, have the same customs, and are affected alike by moral and political influences; and, finally, the tract in which the disease prevails is the richest and most fertile portion of the province.

The natives themselves impute to the quality of waters a powerful influence over their state of health; and when it is recollected that water and farinaceous vegetables constitute the chief diet of Hindoos, any impurity of that fluid would produce effects more readily upon them, than on persons whose food and habits are less simple; but whether they are right or wrong in ascribing the prevalence of goitre to the impurity of particular waters, I shall not at present stop to inquire. A subject on which so many conflicting opinions exist, requires to be elucidated by such facts as, from their number, force, and simplicity, can lead to no erroneous interpretation; and in collecting these facts, the method I adopted, on observing the prevalence of the disorder in one great section of the district, and its absence in another, was to mark the physical characters by which these places were distinguished from each other. The consequence was, a perfect agreement in external aspect, altitude, and climate, but a very marked difference in their geological structure; and this distinction was even traced down to the very villages in which the disease is found, with such perfect nicety as to enable one almost to pronounce, *à priori*, on examining the rocks of a neighbourhood, whether the inhabitants of it are affected with goitre or not.

In pursuing the inquiry farther, it is found that every village is not equally affected in the same neighbourhood, but that some are quite exempt, and others affected to the extent of half their population; and this difference is not found to depend on any accidental or transitory cause, such as usually influence epidemic complaints, but has always affected the inhabitants of a particular village, while those of adjoining hamlets have continued perfectly and permanently free from the complaint.

That this does not altogether depend on hereditary predisposition, is rendered certain, by the numerous cases of persons who, having changed from a healthy to an unhealthy village, have become the subjects of the disease, and from the tumours of those affected becoming stationary, and even disappearing entirely, during a residence in a healthy village. The following details of facts, on which the foregoing statements are founded, will not, I trust (although they are tedious), be thought unworthy of attentive perusal. In order that we may proceed on some fixed and sure principles, I shall not include in the statistical notices any village that has not been inhabited for at least nineteen years, or since the period at which the province fell into the hands of the British. For the same reason, I shall also exclude the villages adjoining and connected with the military posts. To prevent the confusion of names, I shall distinguish the various groups of villages to be noticed alphabetically.

A. Villages of Rykote and Patan, six in number: they are situated two miles north of the military post of Lohooghat. The Patan villages are erected on the southern foot of a lofty mountain, and those called Rykote are surrounded by mountains distinguished by the same name, which ascend from 1,000 to 1,500 feet above them. The villages are elevated about 6,000 feet above the sea, and are erected on a ferruginous clay, derived from the disintegration of clay slate. Their mean annual temperature is about 60°. They contain 200 inhabitants, all of whom are free from goitre.

B. Nine villages situated on the north-west acclivity of the mountain called Gome-dace: their names are, Nakote, Gourouly, Choka, Pimtolly, Jata, Borinkora, Neltokora, Chopota, and Seiligna. They contain 800 inhabitants. Four cases of small goitre only have been found among them, and the four persons are aged. They informed me they acquired the disease in their youth, while residing in a distant part of the country. These villages are erected on clay slates. Mean altitude, about 4,300 feet; mean annual temperature, about 64°.

C. Four small villages on the eastern declivity of Gome-dace. They contain 150 inhabitants, who are all free from goitre. These villages are erected on clay slate. Mean altitude, 3,800 feet. The situation of these villages in the valley of the Mahi Kalee river, renders their temperature high, particularly as they are sheltered from all but the north and south winds.

D. Villages of Pansall, Cheemrouly, Konera, Leno, Simela, Chomonnee, Lund, Katee, and Katully, nine in number, which contain 800 inhabitants. They are situated on low ridges and deep ravines. Mean altitude, about 3,700 feet; mean temperature, 70°. They are erected on primitive clay slate. Only three cases of goitre could be found amongst their population.

E. Villages of Agee, Nina, Choura, Rye, Deortola, Gorong, Sutura, Chakora, Wallishone, and a few others, contain 600 persons, amongst whom, four persons are affected. These villages are erected on primitive clay slate. Mean altitude, about 4,000 feet.

F. Villages of Bentally, Jarig, Babra, Goumana, Biouly, Guinora, Kakur, and Barakote, situated in deep valleys, and on high ridges and declivities. They contain 500 inhabitants, six of whom have goitre. They are erected on clay slate and silicious sandstone. Their altitude varies from about 2,500 to 6,500 feet.

G. Jeercoonee is the name of a mountain which is given off from the Sooee group, about four miles N. W. of Lohooghat. It extends into the valley of the Ponar river on the west; is exceedingly rugged, and, for the most part, inaccessible. Deep ravines enclose it on three sides: in these are situated, on the north, three villages, which are named from the mountain; on the south are the villages of Junera, Dootee, Thur, and Mura: in all, seven. They contain about 400 inhabitants, without a single case of goitre. Nothing can be more alpine than the localities of these villages; the lofty cliffs and mountains seem to overhang them, while deep chasms lie at angles of 30° below. The altitude of these villages is from about 2,200 to 3,500 feet, and their mean temperature about 70°. They are erected on clay slate and silicious sandstone.

H. Rigong, Chimtouly, Bursolly, and Popoulee. The two first are erected on low ridges of clay slate, surrounded by deep ravines and extensive forests; but the last is erected near the summit of a mountain of hornblende slate. These villages contain 250 inhabitants, of whom none are affected with goitre.

Abstract.

Purgunas, or Districts.	Number of high-caste inhabitants.	Number of low-caste inhabitants.	Rocks in which the springs are situated.	Number of high-caste inhabitants affected with Goitre.	Number of low-caste inhabitants affected with Goitre.	Rocks composing the sites of the villages.
A	200	0	Clay slate	0	0	Clay slate.
B	850	100	Clay slate	3	1	Clay slate.
C	700	100	Clay slate	3	0	Clay slate.
D	550	50	Clay slate	4	0	Clay slate.
E	460	40	Clay slate	4	2	Clay slate.
F	200	0	Silicious sandstone	0	0	Silicious sandstone.
	150	50	Clay slate	0	0	Clay slate.
G	50	4	Hornblende slate .	0	0	Hornblende slate.
	170	26	Clay slate	0	0	Clay slate.
	3,330	370		14	3	

NOTE.—To the abstract might be added a population of about 4,000, who inhabit the gneiss and granite district, and who reside a few months of the year in the plains. These persons are affected with goitre in the proportion of 1 in 500, which would make the whole population of the mountains on the south of the Ramessa river 7,700 souls; of these about 25 are affected with goitre, or about 1 in 308.

The foregoing table exhibits the result of the details we have gone over thus far, in a way that will be convenient for comparison with the result of similar inquiries in other portions of the province about to be detailed.

INHABITANTS OF SHORE VALLEY.

The general situation of the district of Shore, as well as its geological structure, has been described. I shall now proceed to notice the villages which are situated in it, specifying the number of inhabitants in each, and their circumstances in relation to goitre; but from the great extent to which the disease here occurs, it will be necessary to attend to more minute particulars than were above observed, in order that nothing may escape calculated to aid the inquiry.

1. A village, called Beesty, is situated on the southern extremity of a low group of hills, in the centre of the valley of Shore. It is erected on clay slate, and supplied with water from the same rock. It contains 60 inhabitants, all of whom are free from goitre, except an elderly person, who came some years ago from another part of the neighbourhood, and the tumour has rather diminished than increased during a residence in this village.

2. Panda is half a mile north from the last described, and is erected on clay slate; but scattered blocks of limestone partly cover the surface of the slate, and a small bed of the limestone terminates in a knoll, on which the village is erected. It contains 25 inhabitants, and all are free from goitre. It is supplied with water from clay slate, and is elevated about 100 feet above the level of the valley. The inhabitants of this and the last village belong to the caste of Rajpoots.

3. Salmora, the name of a few huts situated at the northern extremity of the valley, about a mile from the last-described village. It is inhabited by two families of distinct castes, seven persons in each family. Four individuals of the caste of Domes and two of the Rajpoots have goitre—in all, six out of 14 individuals. This, like the former village, is elevated a little above the valley, and differs from it only in being erected on limestone, from which rock it is supplied with water. One of the Rajpoots is deaf and dumb, with a

large head and idiotic expression of countenance, which are all symptoms of cretinism.

4. Celouly, a small village situated on the northern acclivity of the same knolls on which the two first hamlets are erected. It is elevated 300 feet above the valley. Its altitude and aspect render its temperature less than that of any of the former villages. It is erected on clay slate, from which it is supplied with water. It contains 25 Rajpoot inhabitants, all of whom are free from goitre.

5. Murh, a larger village than any of the foregoing. It is situated close to No. 4, but on the base of an opposite mountain, and contains 70 inhabitants, who belong equally to the castes of Rajpoots and Domes; two of the former, and three of the latter are affected with goitre. It is erected on the junction of clay slate and limestone, and an extensive bed of granitine, composed of felspar, calc spar, and steatite, advances close to the village. The basins of the springs are situated in this rock, but the waters are most likely derived from the substratum, as the granitine in question is seldom observed to afford springs.

6. Goseragong is situated a mile N.N.E. of Murh, and contains 18 inhabitants, 10 of whom are Rajpoots, and the rest are Domes; seven of the former and two of the latter have enormous goitres. The inhabitants of this village assured me that they seldom exceeded the age of 50, being generally cut off by this disease; and a person died of it only a few days before my first visit. The village is erected on a coarse conglomerate of calc tuff, enclosing rolled masses. Lofty precipices of Alpine limestone ascend abruptly behind it, to the height of 2,000 feet; but the site is not more Alpine than that of the seven villages at Jeercoonee (G). Goseragong is elevated about 200 feet above the level of Shore Valley, and has an open aspect only on the S.E.; its temperature is consequently higher than that of the generality of villages in its vicinity; but in none of these respects is it more objectionable than either of the villages of Jeercoonee.

Water is supplied for the use of this village by a fountain, which issues from the limestone rock.

7. Teebee. This village is situated three-quarters of a mile from No. 6, on an arm of the same mountain which extends into the valley. It contains 20 inhabitants, seven of

whom have large goitres. It is erected on clay slate, coated with an incrustation of calcareous matter, and is elevated about 200 feet above the level of the valley, and is watered by a stream which issues from the same source as that which supplies No. 6; but, in the present case, it is taken for use after running about a quarter of a mile in a natural channel.

8. Satgalinge is erected on an arm of a mountain which extends the same distance into the valley as the last. It has the same elevation, and is erected on the same rock, and the localities of both are close together, with a precisely similar aspect. It contains 40 inhabitants, only two of whom have goitres. It is supplied with water from a spring in clay slate. The inhabitants are Domes.

9. Panère — the name by which three small groups of houses are distinguished in the north-western extremity of the valley, at an elevation of about 200 feet above it. Two of these groups contain 30 inhabitants, four of whom have goitre. They use the water of a stream that descends a short way from its source in the limestone cap of the mountain above them. The third portion of the group contains 24 persons, and is furnished with water from a spring in clay slate. They belong to the Rajpoot caste, and are free from goitre. The 30 persons first mentioned are Domes. Panère is built upon clay slate.

10. Bajeetee is erected on the southern acclivity of a low ridge which intersects the valley from east to west. It contains 20 Brahmins and 30 Domes; of the former, three, and of the làtter, 13 are affected with goitre. These people derive the waters which they use for all purposes from two distinct wells sunk near the bed of a stream which issues from the limestone cap of the adjoining mountain. The wells are so situated with regard to the stream, that they do not seem to be supplied by any independent spring, but rather from the stream, particularly during the dry season. The rock surrounding the wells, as well as that on which the village is erected, is clay slate, coated with calcareous matter, especially in all fissures, exposed surfaces, and rifts of the strata.

11. Popdeon is situated half a mile west of No. 10, and contains 80 inhabitants, 50 of the higher and 30 of the lower caste. Of the former, eight, and of the latter, 10 are affected.

There are no distinctions in the physical condition of the inhabitants of this as compared with the last village.

12. Panorah is a village which is situated in the western extremity of Shore Valley, about a mile west of the last. It contains 70 high-caste inhabitants and 20 Domes; of the former, one only is affected with goitre, while six of the latter have large tumours. One of these swellings, which was the largest I had an opportunity of seeing, measured two feet one inch round the neck, and one foot eleven inches from one angle of the under jaw to the other on the opposite side.

The Brahmin or high-caste inhabitants of this village derive their water from a spring in clay slate; and as the prejudice of the Hindoos denies to Domes the privilege of partaking of the water of the same spring, the excluded caste are forced in this, as in many other cases in Kemaon, to use this fluid from what they, as well as the Brahmins, believe to be impure sources. In this instance it is taken from a stream that issues from the same limestone caps that afford water to the two last-described villages. Panorah is built on clay slate, slightly coated with calcareous matter.

13. Paruree. This village is lower down in the valley than any of the foregoing, and is about a quarter of a mile south of the small knoll on which Bajeetee is erected. It contains 60 inhabitants of the Brahmin caste, and there is no case of goitre among them. This village is erected on clay slate, and surrounded by fine springs in the same rock.

14. Dungaunee contains 25 inhabitants. They are free from goitre. This village is situated half a mile south of the military cantonments, on the southern side of the low ridge which intersects the valley from west to east. It is supplied with water from a spring in clay slate.

15. Bagalla is a village situated a little lower in the valley than the last. It contains 18 persons of the higher, and 22 of the lower caste: none of them have goitre. It is surrounded by springs in clay slate, and erected on the same rock.

16. Kumora is situated a quarter of a mile east of Dungaunee, and contains 70 inhabitants, of whom two are slightly affected with goitre. This village is elevated 50 feet above the valley, on a small knoll forming part of the low ridge that intersects the valley, and which has been before mentioned;

the ridge rises behind this, and the following seven villages to the height of about 200 feet above them, at a distance of 300 yards in their rear. This and the following villages are erected on clay slate, and plentifully supplied with water from springs in that rock. The first three in the list are situated close to the limestone, which forms in tabular masses, the rugged cap of the ridge.

17. (*a*) Jakane, 30 inhabitants; all free from goitre.
18. (*b*) Chouser, 50 inhabitants; no case of goitre.
19. (*c*) Beera, 40 inhabitants; no case of goitre.
20. Boorikote, 100 inhabitants; no goitre.
21. Kaseena, 15 inhabitants; no goitre.
22. Kosooly, 25 inhabitants; no goitre.
23. Lailure, 100 inhabitants; no case of goitre.

It is a remarkable fact in the history of this disease in Kemaon, that as far as we have yet proceeded, it appears to extend in lines parallel to the direction of the strata. This important observation is strongly indicative of the influence of particular rocks on the remote cause of goitre. The villages which are exempt from the morbid influence, are those which lie in a line along the base of the central ridge of the valley, beginning with Paruree (13) and extending eastward to Lailure (23) and Murakote; on each side of these are the villages which are affected. They consist of two groups: one a mile distant, in the south-eastern portion of the valley; and the other the same distance in an opposite direction; and both are disposed in lines parallel to the line of healthy villages we have just noticed, as well as to the direction of the strata.

The locality of the three following villages in the south-eastern portion of the valley, bears a striking resemblance, in external appearance and geological structure, to the site of those villages marked *a*, *b*, *c*, whose inhabitants may be said to be insulated from the limestone strata, inasmuch as they derive their supply of water from springs in another rock.

25. (*d*) Kutkora, 15 inhabitants; five goitres.
26. (*e*) Baldakote, 14 inhabitants; seven goitres.
27. (*f*) Batuda, 16 inhabitants; ten goitres.

These three villages are those which compose the south-western line, and are erected on a conglomerate of calc tuff, inclosing fragments of clay slate and other rocks, and partly on clay slate coated with calc tuff. A clay slate mountain, supporting a massive cap of compact limestone, ascends to an altitude of 300 feet above them, pouring out numerous fountains, from which the immense beds of calc tuff, on which the villages stand, have been derived. The waters are beautifully clear and limpid, and are taken for use as they jet from the rock. The peculiarity in the rocks from which their waters are derived, are the only relations in which the inhabitants of the villages *d*, *e*, *f*, differ from those of *a*, *b*, *c*: the altitude, aspect, temperature, religion, and morals of the inhabitants of both groups of villages being the same.

The villages on the north-eastern extremity of Shore Valley, whose inhabitants are affected with the disease, are the following:—

28. Deota, a lengthened village which occupies half a mile of the foot of Durge mountain. One extremity of it is inhabited by Brahmins, the other by Rajpoots and Domes. Of the first caste, there are about 20 persons, all of whom are free from goitre; of the second there are 40, of whom two-thirds are affected more or less; and of the third caste, nearly the whole are affected, 40 in number: so that, including the Brahmins, there are only about 40 persons in this village exempt from goitre, out of a population of 100. To what cause can we ascribe the immunity of one caste of the inhabitants of this village, and the almost entire affection of the other two castes? They are all alike well fed, and have little toil; their land producing the requisites of life almost without labour. Difference of caste does not here imply a difference in pecuniary circumstances, and consequently of the comforts of life. In these respects, the three castes in this village are on a perfect equality; nor will hereditary predisposition, acquired by intermarriages between affected parties, be sufficient to explain the interesting fact: for the affected parties are confined to the castes of Rajpoots and Domes, who cannot intermarry; while the Brahmins and Rajpoots may.

The village is raised about 100 feet above the level of the valley, and the mountain, at the foot of which it is situated, rises with a gentle slope, and is not in this vicinity at all rugged. It is chiefly composed of compact limestone; and the village is erected on a conglomerated rock, composed of calc tuff inclosing masses and fragments of other rocks. There is a spring situated in the valley, at the distance of about 100 yards from the village, which from its first appearance has the character of a mineral spring. The water bursts forth with strong ebullition from numerous veins, in the quantity of at least 40 gallons a minute, and communicates adhesive properties to the sand and gravel by which it is surrounded. The temperature and quantity of the water is the same at all seasons.

The former inhabitants of this village, aware perhaps of the noxious effects of the spring, had an aqueduct formed, by which water is conveyed into the Brahmin portion of the village, from a distant source. The aqueduct being allowed to go out of repair, the quantity of water it transmits is reserved exclusively for the Brahmins; but during the rainy season, when water is plentiful, the Rajpoots also use the water of the aqueduct; but the Domes have no alternative at any season but to use the water from the spring.

The circumstances of this village, with respect to goitre, might of themselves be sufficient to confirm the doctrine of mineral waters; but so much difference of opinion has hitherto prevailed on this subject, that it is not likely such evidence alone can prove satisfactory: more especially as we are in the habit of overlooking occasional facts, as decisive as the above, although they have been brought to light by philosophers, whose names would have been sufficient security for their accuracy on any other question.

29. Ninee. This village is also erected on the foot of Durge mountain, within a mile of the last-described village. It contains 80 inhabitants; there is not one case of goitre among them. These people belong to the Rajpoot caste; their village is erected on clay slate, which is partly detached from the base of the mountain by a small ravine, only a few yards wide. The mountain is here rugged and inaccessible; the village is supplied by a small but sufficient spring, in clay slate.

30. Chonda; also on the foot of Durge mountain, about two miles east of Deota. Chonda is built on a conglomerate of calc tuff, enclosing fragments of slate and limestone. The inhabitants use the water of a stream which descends from the declivity of the mountain, depositing calc tuff. For convenience, a few years ago, a portion of the stream was conducted in an artificial channel, through cultivated land, to the

village—a circumstance which, there is some reason to believe, had modified the effect of the water, for the tumours of those affected have not increased latterly as they used to do, and children continue free from the complaint. Of 25 inhabitants, seven are affected.

31. Sunn. A little village, about a mile eastward of Chonda, on the opposite side of the ridge on which Lailure is erected. It is inhabited by a family of Brahmins, 10 in number, five of whom have goitre. This village is also watered by a stream which descends from the same limestone mountain declivity.

32. Oliel and Cubulcola. Two small hamlets, situated three miles eastward of Sun, in the direction of the strata. These two villages are situated in a most pleasing amphitheatre, completely sheltered from northerly and westerly winds, and partly also from those of the south, but exposed to the full power of the sun, until a few hours before it sets, when the valley is left in shade. There are 25 inhabitants in these villages, 13 of whom have goitre, and 10 of them are cretins; of these a whole family is deaf and dumb. Their deafness appears to depend more on a general insensibility of the mind to external impressions, than on any morbid condition or fault of the ears. They seemed also to be deficient in sight, and quite insusceptible of the passions of joy and fear. The mountains around the locality of these villages are composed chiefly of limestone.

33. Bagultolly lies in continuation of the same line, and two miles east of the last-described, and probably about 1,500 feet below the level of Shore Valley, in what now may be called the valley of the Mahi Kalee. Its aspect and locality are confined by mountains, which, from this low situation, seem to be of great height. It is erected on clay slate, and partly watered by a spring in that rock, and partly by a stream from the mountain. It contains 25 inhabitants, four of whom have goitre. The little arable ground around this village is in a high state of cultivation.

34. Bescolly. This village is on nearly the same level with the last-described, but instead of being in an enclosed valley, it is situated on an exposed though low ridge, composed of clay slate, covered by the usual calcareous conglomerate, and watered by fountains which are poured from the

mountain declivities composed of compact limestone. Of 25 inhabitants in this village, 10 of them are affected with goitre.

35. Gooraght, situated two miles N. E. of the last-described. It is built partly on clay slate, and partly on the conglomerate, which now contains, in addition to the usual rocks, blocks of common serpentine. Water is procured from the mountain declivity. Of 24 inhabitants, 10 have goitre, and a father and two sons are cretins; the sons are both deaf and dumb.

The two villages which we have noticed last, are less important in point of evidence, as the inhabitants, at certain seasons, are compelled to retire to some neighbouring locality, in consequence of the rapacity of wild beasts. I have, however, added them, to complete the account of the population in this direction. For a similar reason, I may add the following villages, which are situated in the eastern extremity of Shore Valley. They are permanently occupied, and are a continuation of the line of villages that are exempt from goitre. The three first are erected on clay slate.

36. Chupuckea, 40 inhabitants, no goitre.

37. Suakote, 40 inhabitants, no goitre.

38. Murakote, 40 inhabitants, no goitre.

Some blocks of overlying limestone are strewed about the neighbourhood of these villages, but clay slate affords a plentiful supply of water for their use.

39. In the lowest part of the valley, where the different streams have collected into a river, which escapes through a deep chasm in the mountains, a patch of silicious sandstone occurs, on which the following two villages are erected:—

(*a*) Kotilla, 50 inhabitants, no goitre.

(*b*) Ruena, 50 inhabitants, no goitre.

40. Deorcolla and Dingas are situated lower down the valley of the small river just mentioned; they are erected partly on an earthy limestone, and partly on clay slate. They are surrounded by many of the highest mountains in Kemaon. The two villages contain 40 inhabitants, and none of them are affected with goitre. The surrounding declivities are overspread with overlying masses of limestone.

Abstract of Shore Valley.

Paragraph under which the villages are described.	Brahmins and Rajpoots.	Domes.	Rocks from which the water is derived for the use of the inhabitants of each village.	Brahmins and Rajpoots affected with Goitre.	Domes affected with Goitre.	Rocks on which the villages are erected.
1	60	..	Clay slate........	..	..	Clay slate.
2	25	..	Clay slate........	..	..	Clay slate.
3	7	7	Compact limestone	2	4	Compact limestone.
4	25	..	Clay slate........	..	..	Clay slate.
5	40	30	Granitine	2	3	Granitine and clay slate.
6	10	8	Compact limestone	7	5	Conglomerate of calc tuff, slate, and limestone.
7	..	20	Compact limestone	..	7	Clay slate, coated with calc tuff.
8	..	40	Clay slate........	..	2	Clay slate.
9	48	6	Limestone and slate	3	1	Clay slate.
10	20	30	Limestone (?)	3	13	Clay slate, incrusted with calc tuff.
11	50	30	Limestone (?)	8	10	Clay slate, incrusted with calc tuff
12	70	20	Limestone and slate	1	6	Clay slate.
13	60	..	Clay slate........	..	..	Clay slate.
14	25	..	Clay slate........	..	..	Clay slate.
15	18	22	Clay slate........	..	..	Clay slate.
16	70	..	Clay slate........	2	..	Clay slate.
17	30	..	Clay slate........	..	..	Clay slate.
18	40	10	Clay slate........	..	..	Clay slate.
19	..	40	Clay slate........	..	..	Clay slate.
20	100	..	Clay slate........	..	..	Clay slate.
21	15	..	Clay slate........	..	..	Clay slate.
22	25	..	Clay slate........	..	..	Clay slate.
23	100	..	Clay slate........	..	..	Clay slate.
25	15	..	Compact limestone	5	..	Clay slate, coated with calc tuff.
26	..	14	Limestone........	..	7	Clay slate, coated with calc tuff.
27	..	16	Compact limestone	..	10	Conglomerate of calc tuff, slate, and limestone.
28	60	40	Limestone (?)	27	33	Conglomerate of calc tuff and fragments of slate.
29	80	..	Clay slate........	..	..	Clay slate.
30	25	..	Limestone........	7	..	Conglomerate of calc tuff and fragments of slate, &c.
31	10	..	Limestone........	5	..	Clay slate (?).
32	10	15	Limestone........	4	9	Clay slate and calc tuff.
33	25	..	Limestone (?)	4	..	Clay slate and calc tuff.
34	25	..	Limestone (?)	10	..	Clay slate and calc tuff.
35	24	..	Limestone (?)	10	..	Conglomerate of calc tuff, slate, and serpentine.
36	40	..	Clay slate........	..	..	Clay slate and scattered blocks of limestone.
37	40	..	Clay slate........	..	..	Clay slate and scattered blocks of limestone.
38	40	..	Clay slate........	..	..	Clay slate.
39	100	..	Silicious sandstone	..	..	Silicious sandstone formation.
40	40	..	Clay slate........	..	..	Clay slate and magnesian limestone.
	1,372	348		100	110	

INHABITANTS OF VARIOUS VALLEYS AND DISTRICTS.

After a few notices of the neighbouring district which surrounds Shore Valley, we may then proceed to the analysis of the whole of the details, and bring them forward in a more interesting shape than we have yet been enabled to do. As far as the nature of the subject will admit, the remaining details may be shortened, by generalizing the villages with the valleys in which they are situated.

41. The great valley of the Ramgungah river is situated eight miles west of Shore. It is here about 1,800 feet above the level of the sea; the lowest villages are about 400 feet above the bed of the river. The mountain declivities on each side ascend at angles of about 20° to the height of 3,000 or 4,000 feet. They are composed almost entirely of compact limestone of various kinds; these rocks rest on clay slate, which occasionally crops from under them. The direction of the valley is from north to south, and of the strata from S.E. to N.W.; the limestone consequently crosses the valley obliquely. There are in this valley eight villages, viz.: Bursar, Kuttygong, Tulsar, Sangur, Domera, Chumalloo, and two others. These villages are interspersed through the valley at different altitudes, and are all erected on limestone. They contain 100 inhabitants, chiefly of the Rajpoot caste; 60 of them are affected with goitre. The following are the limestone rocks found to compose this part of the valley:—

(*a*) Compact limestone, which abounds in such quantity as to form the peculiar alpine character of this portion of the valley.

(*b*) Extensive deposits of calcareous tufa, enclosing rolled masses of other rocks.

42. The valley of Kalapany. This valley adjoins the northern extremity of Shore, and extends six or eight miles to the westward, where it falls into the valley of the Ramgungah (41). It is probably one of the lowest inhabited places in Kemaon, and is closely surrounded by mountains, some of which ascend 6,000 feet above the river, at angles subtending from 20° to 30°. It contains few villages, and the presence or absence of goitre is here marked by the same

circumstances as have been observed in Shore Valley. The following are the two most considerable villages, and are most remarkable in the contrast they present to each other with respect to goitre:—

(*a*) Beechelly. Situated in the lowest part of the valley. It contains 70 inhabitants, 30 of whom have goitre. This village is closely surrounded by mountains of limestone, and is erected on an alluvial deposit, formed of the debris of other rocks, cemented loosely with calcareous matter. 50 of these persons are Rajpoots, the remainder are Domes.

(*b*) Reunna. Situated a mile to the eastward of Beechelly; contains 50 inhabitants, only one has goitre.

This village is as much enclosed by surrounding mountains as the last, and is only about 50 feet higher.

It is erected on the side of a knoll of clay slate ; and, having no spring, the inhabitants use the water of the river, which comes from the valley of Barabice. The inhabitants of this village are Rajpoots.

43. The valley of Barabice is situated twelve miles north of the valley of Shore, and is elevated about 4,000 feet above the sea. It is somewhat of an oval shape, extending about four miles from east to west, and two miles from north to south. The eastern extremity of the valley is composed of clay slate containing beds of talc. There are five villages in this end of the valley, which contain a population of 152; and I could not perceive, or learn upon inquiry amongst them, that there was a single case of goitre in these villages.

The western extremity of the valley is bounded by a low ridge, formed of horizontal beds of compact limestone, resting on clay slate. Six villages, which contain 192 inhabitants, are erected on these rocks; and out of this number, 70 persons are affected with goitre: but it must not be supposed that these 70 persons belong equally to the six villages. The following two instances will show the diversity in this respect.

(*a*) Ager. This village contains 50 inhabitants, 40 of whom have large goitres, and 20 of them are cretins. They use the water as it issues from the drift of an old copper mine, situated in the limestone. The people were earnestly requested to discontinue the use of this water in future; and I pointed out a spring at some distance that they might substitute for it, in the full confidence of being benefited by the change.

(*b*) Ducygong, situated within half a mile of Ager, on the same rock, contains the same number of inhabitants, not one of whom are affected by the disease. They use the water of a spring in clay slate, and belong to the Brahmin and Rajpoot castes.

44. The valley of Deodara is situated on the southern side of Shore Valley, from which it is only divided by a detached mountain of clay slate, with a slight deposit of tabular limestone on its summit. The opposite side of the valley is bounded by Takill, a very lofty mountain.

This valley contains two large villages and several smaller ones. It is well watered by numerous springs in clay slate. It contains 250 inhabitants, of whom four persons only are affected with goitre, and these are confined to a small village which contains 20 inhabitants, and which is erected immediately beneath the limestone cap of the mountain which divides the valley from Shore.

45. The valley of Goron is situated on the western side of Shore Valley, from which it is separated by a considerable ridge. It contains seven villages, and 179 inhabitants; 16 of whom have goitre, and nine of these affected persons belong to a small village, of 24 inhabitants, called Majara: the inhabitants of this village derive their water from a stream which rises in the limestone of the Oudepore mountains, on the south side of the valley.

This valley is composed of a variety of rocks; but the villages are generally erected on clay slate, which abounds in springs of pure water.

A village in this valley, called Chana, is erected on the same granitine as that which occurs at Murh (5). It contains 30 inhabitants, two of whom are affected with goitre.

46. The valley of Roilputty extends along the south-west foot of Takill, and is about seven miles distant from the valley of Shore. It is extremely wild and alpine, and contains only two villages, 25 persons in each.

1st. Tomilly is erected on clay slate, which here contains no water: this is, consequently, procured from a stream which falls in a small but picturesque cascade over the rocky precipices of limestone here forming the declivity of Takill. Six cases of goitre are found in this village, and a third of the whole of the inhabitants approach nearly to the condition of cretins.

2nd. Kurkolly, the second village, is erected on the same rock with the first, but at a lower and more distant portion of the valley. It is furnished with water from the same rivulet, after it has run about a mile and a half along the valley. In this village there is but one case of goitre.

47. The valley of Beechar is connected with the south-western extremity of Shore Valley, and is only divided from it by a low narrow ridge. The water-shed of the valley descends from the north-west, and is composed of clay slate: on this is situated a village containing 40 inhabitants, all of whom are free from goitre. On the lower side of the valley, and near the base of a lofty mountain, are two villages erected on a knoll of argillaceous slate, the surface of which is coated with a slight calcareous incrustation. One is occupied by Brahmins, the other by Domes—about 20 of each sect. Five of the Brahmins and ten of the Domes are affected with goitre. Some of the tumours are extremely large, even in persons of the age of ten years. Water is here afforded by two springs, situated close together. They resemble the spring at Deota (28) in almost every particular. The waters boil up in such quantity as at once to occasion a considerable stream. They were surrounded by so much sand and gravel, that I was unable to ascertain the rock from which they emerge; but limestone is found within a short distance of the springs on the one side, and clay slate on the other, so that the position of the veins from which the waters issue may be presumed to be situated between these two rocks.

Abstract.

Names of the Valleys.	Brahmins and Rajpoots.	Domes.	Rocks from which the water is derived for the use of the inhabitants of each village.	Brahmins and Rajpoots affected with Goitre.	Domes affected with Goitre.	Rocks on which each village is erected.
41. Ramgungah	100	0	Limestone	60	0	Limestone.
42. Kalapany { *a*.......	50	20	Limestone	20	10	Conglomerate of calc tuff,&c.
42. Kalapany { *b*.......	50	0	River....	1	0	Clay slate.
43. Barabice { E.......	120	32	Clay slate	0	0	Clay slate.
43. Barabice { W......	110	82	Limestone	30	40	Limestone.
44. Valley of Deodara..	145	80	Clay slate	0	0	Clay slate.
	25	0	Limestone	4	0	Clay slate.
45. Valley of Goron	0	24	Limestone	0	9	Limestone.
	90	35	Clay slate	5	0	Clay slate.
	30	0	Granitine	2	0	Granitine.
46. Valley of Roilputty ..	25	25	Limestone	1	6	Clay slate.
47. Valley of Beechar	40	0	Clay slate	0	0	Clay slate.
	20	20	Limestone	5	10	Clay slate, incrusted with calc tuff.
	805	318		128	75	

RESULTS ATTAINED FROM THE FOREGOING.

Names of Rocks.	Number of villages.	Number of inhabitants.	Number of persons affected with Goitre.	Number of Cretins.	Mean altitude.	Mean temperature by Fahr.
1. Granite and gneiss (*see* Note, p. 67)	0	0	0	0	6,500	68°
2. Hornblende slate and mica slate	1	50	0	0	6,000	..
3. Clay slate	71	3,957	22	0	4,100	78
4. Green sandstone	3	200	0	0	3,500	..
5. Granitine	2	100	7	0	4,000	..
6. Silicious sandstone	1	40	0	0	..	..
7. Alpine limestone	35	1,160	390	34	4,000	78
	126	6,543	430	34		

48. From the above figured abstract, it appears that the proportion of the inhabitants of each rock, affected with goitre and cretinism, will stand to the healthy in the following order:—

Granite and gneiss—Goitre, $\frac{1}{500}$; cretins, none.

Mica slate and hornblende slate—Goitre, none; cretins, none.

Clay slate—Goitre, $\frac{2}{385}$; cretins, none.

Green sandstone—Goitre, none; cretins, none.

Calcareous rocks—Goitre, $\frac{1}{3}$; cretins, $\frac{1}{32}$.

Are we to suppose that these results are the effects of chance, or of an accidental association of circumstances confined to a particular spot? When we recollect that a space of upwards of a thousand square miles has been made subject to the inquiry, and that, in every portion of this space, the same invariable circumstances attended the presence of the disease, and that its absence was invariably marked by the absence of those circumstances, it is impossible not to view them in the light of cause and effect.

DESCRIPTION OF GOITRE AS IT OCCURS IN KEMAON, WITH AN ATTEMPT TO ACCOUNT FOR THE FOREGOING RESULTS.

49. A mistaken notion appears to be entertained by some authors, who speak of the comparative innocence of the disease, and consider its unsightly appearance as its worst effects; thus depriving the inquiry of that interest which is felt in the prosecution of researches into the nature of other disorders. Those who express themselves so, cannot have had sufficient opportunities of witnessing the misery entailed on the inhabitants of tracts of country in which the goitre prevails. There is no disease of which the people have greater dread, or from which they are more anxious to be relieved. It is true, there may be little acute pain at first; the ultimate effects of the disorder are however, scarcely less fatal than those of any other complaint to which we are subject; to say nothing of its complication with cretinism, the greatest of all afflictions.

The first object, is to inquire whether the goitre of Kemaon be a peculiar disease: this point will be best determined by a brief description of the nature of the complaint.

The tumour does not always originate in the thyroid gland; but in a third of the cases I have seen, it appeared to commence with a fulness of the base of the neck, on one or both sides, over the middle of the clavicle; from thence the swellings ascend, and, in a longer or shorter time, reach the situation of the thyroid gland, when both tumours unite.

In its progress up the neck, the tumour sometimes appears to become entangled, as it were, in the folds of the fascia cervicalis; it then becomes indurated, and forced between the trachea and the œsophagus, where it displaces these organs, and often proves fatal by interrupting their functions, before it increases to any considerable size, and without even extending to the thyroid gland, being apparently confined to the lymphatics which accompany the great vessels of the neck.

In its more common form, the thyroid gland is first affected, and the tumour increases to a great size, without causing much inconvenience, especially if it be loose and pendulous; but in many cases of this form of the disease, the tumour is probably compressed in the way above described, and dyspnœa becomes the most prominent symptom; the lips become darker than natural, the eyes blood-shot and protuberant, and the patient dies from protracted strangulation.

The disease begins at any period of life after the age of three years, and never, as far as I have seen, arrives at its full size sooner than six years from the time of its commencement, but is generally much slower; its progressive augmentation seldom however, becoming perfectly suspended during a residence in an affected village.

This description must be received as in some measure empirical, the prejudices of the Hindoos of Kemaon being against the dissection of their dead.

The usual size of a full-grown goitre is about one foot ten inches in circumference, including the neck; and about two feet from one angle of the lower jaw to the other of the opposite side (measuring under the tumour).

Incipient tumours of only a few months' or a year's duration are easily dispersed by stimulating linaments, and a few alterative doses of calomel; but without the change of the accustomed water, these means will only afford temporary relief.

50. From the above description, there can be no reason to doubt the identity of this disorder with the strumous endemic of Switzerland; and wherever it is found, from Abyssinia and the Chinese Wall to Sumatra and Derbyshire, it appears to present the same characters, and is less under the influence of general causes than, perhaps, any other complaint. Does not this fact establish the importance of such inquiries, as opening a new field to our researches into the nature of endemic contagion?

The next point is to inquire into the manner in which the results in question are affected, by what is generally known respecting the physical structure of those countries in which the disease is endemic.

51. From the writings of geologists, we learn that alpine limestone does not occur to any great extent in the mountains of Ireland, nor in those of Scotland and Wales; and in these countries, goitre is unknown. In England, the disease is known by the name of the Derbyshire neck, and is principally

confined to Derbyshire, where the particular rock in question forms the characteristic features of the country.*

In the Alps of Switzerland and Tyrol, where goitre and cretinism both prevail, we have the authority of geologists, that alpine limestone and nagelflugh compose the greatest portion of the mountains. Humboldt mentions that nagelflugh covers the greatest part of Switzerland, to the height of a thousand toises. Now this nagelflugh† is the same rock, or nearly so, as that on which the villages of Goseragong (6), Batuda (27), Deota (28), and Chonda (30), are erected, villages whose inhabitants are affected with goitre to the extent of half their population. This rock is composed in Kemaon of a basis of calc tuff, inclosing fragments of other rocks, from the size of a grain of sand to that of a millstone. These fragments are either rounded or angular, and the basis in which they are imbedded is either solid or vesicular. The matrix of the rock is a chemical deposit derived from water, and the inclosed masses which it contains appear, at first view, to indicate that they were broken and precipitated to their present situation. By more attention to the changes going on in nature, the formation of nagelflugh appears to be extended down to our own time, as an alluvial deposit, occasioned by the constant crumbling of rocks, and rolling down of masses and fragments separated by the chemi-

* In Cumberland and Wales there are more lofty mountains than in any other part of England, Scorfell and Snowdon being nearly 3,300 feet above the sea; while Axe-edge, the highest peak in Derbyshire, taken by Colonel Mudge, is only 1,751 feet: so that goitre in the latter country cannot be owing to the height and magnitude of its mountains, but the cause must be sought in their structure; accordingly, we find the mountains of Derbyshire are composed of alpine limestone, while those of Cumberland and Wales consist, for the most part, of granite, clay slate, porphyry, and sienite.

At Fribourg, Valteline, Berne, Pay-de-Vaux, Dresden, Savoy, and Piedmont, the most remarkable districts in Europe for the prevalence of goitre, alpine limestone constitutes the principal rock formation.

† "Nagelflugh," says Professor Jameson (*Syst. Min.*, 1808), "is usually composed of fragments of limestone, more or less rounded, and of various magnitudes, cemented together by a basis of calc sinter. It occurs always at a greater or less distance from limestone mountains, and sometimes forms considerable tracts of country." It appears abundantly at the foot of the great hills of alpine limestone that bound Bavaria to the south, and in many other places in the great limestone range that passes through Tyrol, Styria, &c.

cal and mechanical agencies of the atmosphere, and again consolidated by the deposit of calc tuff from the waters of alpine limestone.

The different appearances which a rock of this nature must necessarily assume, has procured for nagelflugh a greater variety of designations than any other formation; calcareous sandstone, breccia, conglomerate, and pudding-stone, are names that have, no doubt, been applied to different varieties of it by English writers. It is best distinguished by being always subordinate to alpine limestone, and it is on this latter account only that its connexion with goitre appears to be important.

52. Alpine, or compact limestone,* does not admit water by percolation, through its solid substance, by means of porous or absorbent qualities; but by open rents and fissures, which communicate with subterraneous caverns in the centre of mountains, where it may either remain for ages, or flow out by counter fissures.†

* It is the "erster floetz-kalkstein," or first floetz-limestone of Werner; the lowest stratum of it is the bituminous marl slate, or the copper slate of the miners.

† Speaking of alpine limestone, Humboldt says (*Pers. Narr.*), "It is the rock that so often interrupts the course of rivers, by engulphing them into its bosom."

"The whole of that enormous mass of limestone at Craven, in Derbyshire, from Ingleborough to Whernside and Gerdal, is intersected by perpendicular fissures, which are narrow at the top, and become wider as they descend, through which the water may be heard at a vast depth below . . . . . Castleton and Poolshole, near Buxton, and Yardas Cave, under Whernside, in Craven, Gerdal, Scar, and Weathercock, in the same district, can scarcely be called caverns, as they are open to the day; but the latter was formerly a cavern, of which the roof has fallen in." "In all these caverns, and others which I observed," says Mr. Bakewell (*Introd. to Geolog.*), "there is a stream of running water; and I am inclined to think, that the caverns have been formed by the agency of water, percolating through fissures; and, in the lapse of ages, excavating the softer or more broken part of the rock."

"The mines seem to be, or to have been, open channels, through which the waters pass within the earth, and, like rivers, have their small branches opening into them in all directions, which are by miners called feeders of the lode. Most mines have streams of water running through them; and when they are found dry, it seems to be owing to the waters having changed their course. . . . . Sometimes the mine is lined with an intermediate substance between the lode and itself: this is the wall of the lode. . . . . The springs in these parts are always hard, as abounding

Water thus circulating through confined caverns, without having undergone previous percolation, is likely to assume changes dependent on the various vegetable and animal impregnations with which it may be loaded; these engender new agencies, which operate on the numerous mineral substances with which the water comes in contact. In the confinement of narrow caves and fissures of limestone, the surface of water becomes so much extended, that every portion of the surrounding rock is exposed to its action. No other rocks contain such extensive repositories of extraneous fossil and metallic substances, and no other formation of rocks contains such extensive caverns and fissures, where these foreign substances are exposed to the slow action of water: hence, the greater number of mineral springs that abound in calcareous, than in any other rocks.*

very much either in stony or sulphuro-saline particles."—*Nicholl's Observ. Nat. Hist. of Mines.*

I might quote farther observations of Dr. Nicholl, one of the most eminent physicians of his time, illustrative of the changes to which water is exposed in the bosom of the earth; but his papers may be consulted in the *Phil. Trans.*

"The three rivers, as they are commonly called, in Peakshole are only some parts of the cave deeper than the rest, and receiving all their waters from the spring, which comes from the farther end of the cave. The waters which pass through Poolshole are impregnated with particles of limestone, and so have incrusted the whole cave in such a manner, that it appears like one solid rock."—*Nat. Hist. Derbyshire, by J. Martyn.*

See also the description of the caves near Bayreuth, in *Phil. Trans.*; also, "Observations on the Nature of Intermitting and Reciprocating Springs" (Atwell, *Phil. Trans.*). "These mountain caverns will account for the statement of Pliny (lib. xxxi. 4), that earthquakes pour out and drink up waters." See account of the great earthquake at Naples, 1731. —*Phil. Trans.*

Mathias Belias describes two caverns in Hungary (*Phil. Trans.*)— "The one emits noxious vapours, and is overflowing with water, which deposits a tophas. . . . . When subterraneous waters flowed from the interior of the fountain in the hidden passages, the ground began to give way, and at length formed a new opening, when it began again to emit noxious vapours, destructive to birds and other animals.

"In the cavern is heard the murmuring noise of running waters, so that a river probably flows through the interior passages, and at last loses itself in some kind of shallow."

* Such as the mineral springs of Buxton, Matlock, Malvern in Derbyshire, Bath, Bristol, and the springs of Imaw in Suabia, Carlsbad in Bohemia, and the salt springs of Konigshorn, and those of Seltzer, Sydchutz, Spa, Pyrmont, and the baths of Carolin in Bohemia, and probably many other celebrated mineral waters, whose physical topography I am unable to refer to at present.

53. Having endeavoured to explain the influence that alpine limestone is capable of exercising on the waters of a district, it remains to offer a few observations on the effect which calcareous rocks may, under peculiar circumstances, exercise on the condition of the air in their vicinity.

The peculiarity of air in mountains has been often brought forward as the exciting cause of goitre, although no attempt has ever been made to explain in what the peculiarity alluded to consists, or why it should exist, further than that it is supposed to be excited by a warm atmosphere, in situations where the free circulation of air is impeded; but now that we have traced the disease to a peculiar constitution of strata, our notions on this intricate point may soon become more precise, and it deserves to be inquired into, whether or not the exhalations from limestone rocks contain a larger proportion of carbonic acid gas than is found to exist in the general atmosphere.

Peculiarities in the physical and chemical constitution of mountain rocks have been hitherto quite overlooked as a source of endemic contagion, which may in some degree perhaps account for the little success that has attended the researches upon this subject; for although it has been known in all ages that there is a difference in the air in different places, by certain effects on the human constitution, yet all that has ever been demonstrated by those who have entered upon the inquiry, was the imperfection of our most refined chemical tests—that, in fact, some further improvements must be made in chemical science before the nature of contagion can be demonstrated.

Humboldt found the proportion of carbonic acid gas in the atmosphere to vary from 0·01 to 0·005 of the bulk of the air; but he does not appear to connect this important variation with local peculiarities of geological structure.

It is well known, that air containing 0·1 of its proportion of this gas extinguishes light, and is speedily destructive of animal life; and as this volatile poison exists in limestone, to the extent of 44 parts in 100 of the solid rock, it is possible to conceive that a sufficient quantity of it, to cause a more or less vitiated condition of the air, may be extricated from limestone by atmospheric heat, assisted by such other causes as promote the decomposition of the rock.

This gas floats on the surface of the earth in places from which it is disengaged; it is evolved by mineral springs, and by all waters which contain it; and it is separated from limestone, the great repository in which it abounds in nature, by *heat:* and the important questions that remain to be decided are, whether the heat of the atmosphere is sufficient to disengage it in any noxious quantity; and whether if, by means of pyrites, assisted by moisture and atmospheric heat, an insensible evolution of carbonic acid gas may not constantly take place in certain localities? These, next to the examination of the waters, are points which are entitled to careful attention.

A reference to the mineral topography of all the villages in Kemaon which I have examined, but one, seems to favour, rather than negative, these views; and even with regard to the village of Ager (43), the occupation of the inhabitants as miners, to which they have been brought up from childhood, may expose them sufficiently to impure air to occasion much of their bodily infirmities, independent of any noxious evolution of gas, in the way we were supposing it possible to occur.

If there be difficulties in the way of conceiving the possibility of the disengagement of carbonic acid gas from limestone, its absorption by lime-water may be suggested as a means by which it may be attracted by the moisture on the surface at the base of calcareous mountains.

The thin incrustation of calcareous matter, so often observed on the surface of clay slate, composing the site of many of the affected villages in Shore Valley and its vicinity, may have been formed by particles of lime having been partially reduced by heat and drought on the adjoining acclivities, and carried by the winds to the knolls of slate, whose moist and absorbent surfaces arrested their further drift, and converted them into a cement, by the absorption of carbonic acid from the general air. Until we are better informed, we should certainly not be too ready to despise the effect which such operations may have on animate, as in inanimate nature; and attention to them might assist in explaining the cause of this disease in certain low tracts extending along the base of the Alps, as well as the Himalayas.

GENERAL OBSERVATIONS ON THE EXAMINATION OF WATERS.

54. In the examination of the waters of a province, or extensive district, it may be necessary to keep in view the geological distinctions by which the several portions of it are characterised, as any peculiarities in the qualities of waters, when derived from springs in the earth, must depend on the nature of the mineral substances which compose their localities. The importance of this observation has always been known, and generally attended to; but it may have derived additional consequence from what has transpired in the preceding sections.

Water constitutes the medium by which living bodies are supplied with new materials. Plants will not vegetate without it, and most of those earthy, alkaline, and metallic substances, which are common to spring-waters in general, have been found to constitute a portion of vegetable bodies. There is no reason to suppose that these substances are secreted in plants by any property of the living principle; we must therefore believe them to be imbibed from the soil, the requisite qualities of which must vary with the habit of species.

The same observations apply to animals; but as they are endowed with more complicated functions, they may be supposed to be still more sensitive to the influence of external agents, according to the rocks and soils they inhabit. The presence of lime, sulphur, magnesia, silica, iron, and manganese, in animal bodies, has been long known, although their production is supposed to be incompatible with the functions of living organs. We are thus led to believe, that these extraneous substances may differ in their proportions in living bodies, according to the peculiarities of geological structure in particular places; as, however, a certain proportion of these foreign matters is essential to the healthy state of animals, so any deviation in this respect may be the cause of disease, as struma, plica polonica, and scorbutus.

Although waters containing impregnations, to a degree causing powerful effects on the human constitution, are found in almost every country, it is not a little surprising that the several degrees of impurity between the medicinal and the pure water should be so much overlooked. The difficulty of pointing out the various impurities, and the unusual circumstance of any being perfectly pure, is, no doubt, the cause of the general indifference of physicians to the quality of water.

It is remarked, that the complete analysis of water is one of the most difficult operations in chemistry. The difficulties, or rather the imperfections, are occasioned by the minuteness of some substances, and the evanescence of others, with which water may be impregnated. The active principle of the Lokarne water was found to escape in four hours through a glass vessel, though corked and sealed in the most careful manner, in a northern climate. Those who deny the influence of particular waters in causing the goitre, merely because they cannot demonstrate the noxious principle and its *modus operandi*, should recollect that although snow is known to form the purest natural water, yet no one would venture to deny the effects on animals ascribed to it merely from its deficiency in common air; and although the noxious effects of hard water has been found to be imaginary, it serves to show the importance attached to certain conditions of water, which differ but slightly from a wholesome standard.

The ancients observed the diminutive size and sickly appearance of plants growing on mountains which contained metals; and although this notion has been lately condemned by some as fanciful, yet we should remember that there is generally something in the appearance of a mineral district to distinguish it from one favourable to agriculture.

I have been led to these remarks by some facts which might tend to show that the noxious principle in the waters of alpine limestone is a subtle combination derived, perhaps, from those strata of the rock which are called by miners "copper slate." They are so distinguished from the quantity of metals which they contain, particularly the ores of copper. In describing the locality of the springs which supply those villages where the inhabitants suffer most from goitre, they may be said to be generally derived frem the strata in question, or, at least, from the lower beds of limestone near the junction where it rests on clay slate.

On the other hand, we were assured by Werner that the water which flows from metallic veins ought to carry along with it a quantity of metallic particles, which, however, it does not; and even in those countries which contain the greatest number of mines, the water rarely contains a small portion of iron—scarcely ever any particles of copper—never silver, lead, tin, zinc, cobalt, mercury, or arsenic. Without dwelling on this apparent paradox, it is sufficient to mention that metals in their mineral state are insensible to chemical tests. Thus, copper pyrites and iron glance, as well as iron pyrites, were reduced to the finest powder and mechanically suspended in distilled water, to which prussiate of potash was added, without affording the slightest indication of their presence. The same was repeated with tincture of galls with the same result.

It was mentioned by Klaproth, that even iron, a metal for which we have such excellent tests, is capable (when in small quantity, as in the Carlsbad waters) of 'eluding the senses, as well as the efficacy of reagents, unless examined at the spring.' Hence it is, from the small quantity in which they occur, and the imperfection of analysis, that certain agents or active powers have not been found in waters, although their presence is known by medicinal effects, as in the case of the Bath waters.

EXAMINATION OF THE WATERS OF KEMAON.

55. The first waters to be examined are those of Goseragong (6), Deota (28), Ager (43), Batuda (27), and Beechar (47).

General accounts of the situation and appearance of these respective springs have already been given; but, to save the trouble of referring back, I may be allowed to repeat what has been said, and to add such further observations as may complete the account of their physical characters.

1. Goseragong water issues from a crevice at the base of a lofty precipice of limestone; the jet is situated at the junction of that rock with clay slate. The quantity of water which issues is about two gallons a minute, and its temperature is 67° Fahr. at all seasons, which is about the mean annual temperature of the place. It is perfectly limpid, and sparkles briskly when poured into a glass. The taste is slightly acid and agreeable; a large quantity of calc tuff is deposited from the stream as it falls down the surface of the mountain. Its specific gravity is 1·009.

2. Deota. The basin of this spring is situated in a low part of the valley of Shore, and is composed of red clay, gravel, and marl, loosely cemented. The spring has a lengthened form, extending in the direction of the neighbouring strata; and if two or three adjoining springs of the same character be included, the whole may be said to be about 300 yards long. The surface of the water is disturbed like a boiling caldron, from the violence with which it issues from the earth. The larger spring discharges enough of water to put a mill in motion. Temperature, 65° Fahr., which it retains at all seasons. Taste, less agreeable than Goseragong water; but still sweet and equally brisk, but less crystalline. After remaining a certain time in corked bottles, it deposits slight bluish-grey flocculi.* Its specific gravity is 1·001.

3. Ager. This water issues from the drift of an old copper mine, in the quantity of about half a gallon a minute: the source of the spring is between the limestone and graphite, which contains copper pyrites. It is perfectly limpid, and sparkles when poured fresh into a glass.

On standing for a few months in bottle, a slight precipitate of ragged putrescent-like matter descends slowly to the bottom.† Its specific gravity is 1·0009.

Batuda. This spring is exactly similar to that of Goseragong in all particulars, the jet holding the same geognostic position, the water itself crystalline, acidulous to the taste, and possessed of the property of forming calc tuff. Its specific gravity is 1·0010.

* These flakes were collected on a filter, dried, and found to weigh about a tenth of a grain from a bottle of water. They were very slightly soluble in nitric acid; the residue, which amounted to very nearly the whole, was melted on charcoal, before the blow-pipe, with borax, to which it gave a yellow colour.

The globule thus afforded, being reduced to powder in a glass mortar, and dissolved in nitric acid, was tested with prussiate of potash, and afforded a precipitate of iron.

† This flocculent matter from a bottle of Ager water was collected and weighed, but though bulky, was too light to make any sensible impression on the balance (not a very delicate one). It was moistened with distilled water, and gave a red tinge to oxymuriate of mercury.

On the application of the white flame of the blow-pipe to this matter, it entirely volatilized, so that it may be considered as of an ammoniacal nature.

5. Beechar. Here we have an almost perfect identity in the appearance of the springs with those of Deota. The water ascends in perpendicular columns, so as at once to form a considerable stream. The source of the spring is situated between the strata of limestone and clay slate, and is concealed by an accumulation of gravel slightly cemented with calcareous tufa. Temperature, 64° Fahr., when the general air in the shade was 82° Fahr., and it never changes: it has consequently the character amongst the natives, in common with the four preceding waters, of being a cold spring in warm weather, and a tepid one in the cold season. The water is clear and very faintly acidulous to the taste. Specific gravity, 1·0011.

56. Effects of precipitants on these waters:—

a. At the spring they very faintly redden litmus, but the change is not permanent.

b. Goseragong water; four hours in bottle, at a temperature of 85° Fahr., but secured as carefully as possible with a common cork, afforded, by the application of lime water, two grains of dry carbonate of lime from three ounces of the water. Beechar, Deota, and Batuda waters, under similar circumstances, each afforded about the same proportion. I was unable to procure Ager water so fresh, and this test was not applied to it until after it remained two months in bottle, when it afforded an opalescent precipitate, which it was scarcely worth while to weigh.

c. Cold infusion of Brazil wood is rendered blue by all these waters.

d. Infusion of turmeric is slightly reddened by those of Beechar, Ager, and Goseragong.

e. Tincture of galls displays slowly, copious flocculent precipitates in all these waters. The colours are, first reddish yellow, changing to greenish yellow, from this to greenish blue, and eventually to black. To complete these changes, a shorter or longer time is required in the different waters: Beechar requires twelve hours, Goseragong fifteen, Deota and Batuda about eighteen, and Ager about twenty-four hours.

f. Prussiate of potash effects no change in any of these waters, except an acid be added; a faint bluish tinge is then produced, and is more conspicuous in Goseragong and Batuda, but sufficiently perceptible in each of the others.

g. Nitrate of barytes occasions no change in the appearance of any of these waters.

h. Oxalic acid causes a slight precipitate in the waters of Goseragong, Deota, Beechar, and Batuda; but Ager water retains its crystalline appearance under this test.

i. Nitrate of silver affords precipitates in all these waters, but the precipitate is redissolved with effervescence in nitric or any acid; and these precipitates are white, except that which is afforded by Goseragong water, which has a tinge of brown. The proportion of these precipitates, from 500 grains of each of the waters, is as follows:—Deota, $\frac{1}{6}$ of a grain; Ager, $\frac{1}{8}$; Goseragong and Beechar each affords about $\frac{1}{12}$; and Batuda about $\frac{1}{14}$ part of a grain.

k. Acetate of lead forms precipitates from each of these waters, and the precipitated lead dissolves with effervescence in acetic acid.

l. Carbonate of potash causes a slight precipitate from Deota water, and carbonate of ammonia produces an ash-coloured cloud from that of Beechar.

Inferences.

1. From experiments *a* and *b*, connected with some of their physical characters, it is evident that disengaged carbonic acid is present, which appears, from *c* and *d*, to be in excess with a base; and, therefore, it requires that it should be present in a quantity equal to one sixth of the bulk of the water, at least, to effect a change of colour in litmus.

2. From experiments *c* and *d*, as well as those of *f*, *g*, *h*, alkalies are evidently present; for, although earthy carbonates or sulphate of lime would change the colour of Brazil wood to blue, it does not appear that these salts are present in sufficient quantity; and, even if they were, they could have no effect on turmeric.

3. From experiments *e* and *f*, iron is the only metal whose presence is clearly indicated.* These precipitates are pro-

* The prussiate of potash, prepared after the simple manner directed by Bergman, indicated a cupreous tinge in the waters of Beechar, Goseragong, and Ager, but I could discover no mineral acid with which copper could be combined; and I knew, by synthetical trials made for the purpose, that particles of copper pyrites, if held in suspension, would yield no

bably modified by the presence of sulphate of lime, although sulphuric acid is not indicated by any of the other experiments.

4. Experiments *g*, *i*, *k*, prove the almost total absence of any of the mineral acids, either combined or free, in any of these waters. The precipitates in *i* and *k* are evidently occasioned by an alkali, assisted probably, in all but Ager water, by a small quantity of lime. A little earthy matter is indicated in Deota and Beechar waters by experiments *l*.

5. From a review of the whole of the foregoing, it appears that the only substances in these waters are, carbonic acid, a small quantity of alkaline matter, a little iron, and lime (except in Ager water), with a scarcely sensible portion of earthy matter in Deota and Beechar waters.

58. Principles of these waters collected by evaporation:—

A wine bottle, or about 13,000* grains, of each of these waters, were evaporated to dryness by solar heat, to which they were daily exposed in such a manner as to guard against accidental impurities falling into them. The following are the quantities of solid extracts from each, when dried at the temperature of 105° Fahr.:—

Ager water, 2 grains, equal to $\frac{1}{6500}$ of the whole.
Beechar ditto, $2\frac{1}{4}$ grains, equal to $\frac{1}{5777}$ of the whole.
Goseragong ditto, 2 grains, equal to $\frac{1}{6500}$ of the whole.
Deota ditto, $1\frac{1}{2}$ grains, equal to $\frac{1}{8666}$ of the whole.
Batuda ditto, 2 grains, equal to $\frac{1}{6500}$ of the whole.

---

characteristic precipitate of copper with this test. Under these circumstances, I applied to the Medical Board for a small quantity of the prussiate of potash used for chemical purposes.

This prussiate of potash I found to be very powerful in discovering the smallest quantity of the salts of copper in solutions; yet it did not confirm the cupreous indication of the other preparation, but I found both preparations equally incapable of detecting the presence of metals in their mineral state.

* A quart of water (wine measure), at 62° Fahr., barometer 30°, weighs 58·443 grains troy; but these waters were exposed to a temperature of above 80° Fahr., and at an elevation of 6,000 feet; so that, in assuming 13,000 grains as the weight of each bottle of water evaporated, we may not be far from the truth. Although a table of equivalents between the measure and the weight of fluids, at different altitudes and temperatures, would be highly useful in such inquiries, here, of course, we do not aim at perfect accuracy, and do not pretend to have suffered any inconvenience from the deficiency here alluded to.

These precipitates consist, first, of a greyish-yellow matter, which was found on the higher margins of the dish in which the evaporation was conducted. It was tasteless, and insoluble in water, but soluble with effervescence in muriatic acid, from which it was precipitated by oxalic acid, and thus proved to be carbonate of lime.

Secondly, below the carbonate of lime were slight metallic precipitates, distinguished by their lustre and iridescent appearance. They were placed on charcoal, and exposed to the white flame of the blow-pipe, when they gave out a stream of air, which carried off a considerable portion of them. The remainder blackened in all but the extract from Ager water, which, with that of Beechar, gives a yellow colour to the glass of borax; while the others render that glass dark green. Patches of metal are seen in the globules thus produced from each of the waters; and these globules, separately reduced to powder in a glass mortar, and dissolved in nitric acid, afford in their solutions, under the application of the prussiate of potash, some a purple, and others a deep Prussian blue precipitate.

The gas which first escapes in this process is, no doubt, carbonic acid, from saline particles of metal, or, probably, from adhering portions of the earthy portion of the precipitates; while the blackness which the imperfect scoriæ assumes is occasioned by the sulphur, contained, no doubt, in particles of pyrites.

The third constituent part of the extract of these waters is found in largest quantity near that portion of the bottom of the vessel from which the last portion of the water ascended, and consisted, as well as I could ascertain, of carbonate of soda, and a small quantity of carbonate of potash, which gave a deliquescent character at first to all the precipitates except that of Batuda.

As to the proportion which each ingredient in these waters bears to the aggregate extract of the same water, it may not be worth while to state, and, perhaps my doing so, might be affecting greater nicety than the means adopted may justify; but as it may afford means of comparison between the results of these and similar inquiries, I mention them—

1. *Ager Water.*
Carb. lime, $\frac{1}{5}$.
Carb. iron, $\frac{1}{5}$.
Carb. soda, $\frac{2}{5}$.
Carb. potash, and a trace of sulph. lime and sulph. } $\frac{1}{5}$.

2. *Goseragong Water.*
Carb. lime, $\frac{1}{3}$.
Carb. iron, $\frac{1}{3}$.
Carb. soda, Carb. potash, Earthy sulphureous, } $\frac{1}{3}$.

3. *Deota Water.*
Carb. lime, $\frac{1}{3}$.
Carb. iron, $\frac{1}{3}$.
Carb. soda, Carb. potash, earthy, } a trace.

4. *Beechar Water.*
Iron, $\frac{1}{3}$.
Carb. soda, $\frac{1}{3}$.
Carb. lime, Carb. potash, Sulph. earthy, } $\frac{1}{3}$.

5. *Batuda Water.*
Carb. lime, Carb. magnes. } $\frac{2}{5}$.
Carb. soda, $\frac{1}{5}$.
Carb. iron, $\frac{1}{5}$.

Beside the above minute quantities of solid ingredients, carbonic acid is also to be considered as present in more than ordinary proportion in these waters. The high atmospheric temperature to which they were exposed, in carelessly-corked bottles, for some time before they were submitted to experiment, as well as the want of any pneumatic apparatus, tend to render the quantity of carbonic acid uncertain; but an estimate may be formed by its effects, even under these unfavourable circumstances, in saturating lime. The quantity of calcareous earth thrown down from lime-water, in the foregoing experiments, must be considerably under what would have been afforded, had proper means been available in order to prevent the dissipation of the gas by the use of stopple bottles at the springs; but those to whom we are to look for the most successful results in such inquiries, must be content with very imperfect means.

59. Having examined the waters of those villages whose inhabitants are most severely affected with goitre, the question is not, whether we have detected any ingredient in them which we can consider capable of causing the disease, but whether these waters differ in any of their characters from spring-waters in general, and more especially from the waters of those springs in their immediate vicinity, which are used by other villages of the same people, but who are free from the complaint. Could such a difference be established, on clear and indisputable grounds, we might then congratulate ourselves on having reached the second tangible step in this interesting inquiry.

The waters now to be examined, for the purpose of comparison, are selected indifferently from those localities whose inhabitants are free from goitre:—

1. The water from the village of Paruree (13).
2. The water from Boorikote (20).
3. The hospital spring at Lohooghat.*
4. The spring situated between the hospital and the lines at Lohooghat.†
5. The spring on the east of the lines at Lohooghat, which is used by the troops at that place.‡
6. The spring on the north of the lines, at the same place, also used by the troops.§

These springs are all situated in clay slate, and the persons who use their waters are perfectly free from goitre.

Respecting springs in clay slate generally, it may be remarked that they are much more numerous than those in limestone, in proportion to the extent of the two rocks, and that they do not appear to be derived from any great depth in the earth.

Their temperature, consequently, falls considerably during the winter, and rises again in summer. They are usually met with in sequestered spots, and covered by dense cupolas of flourishing shrubs. Their waters never rush from the earth with violence, or in greater quantity than to occasion a slight ripple on their surface. They are clear, but rarely very crystalline or sparkling; and in all these circumstances they form a striking contrast with the waters of springs derived from limestone, whose chemical properties are often shown at first sight, by the deposits of tuff at the head of the spring, the want of vegetation, and the ebullition and quantity in which they are thrown up. It would be tiresome, as well as useless, to describe, separately, the physical qualities of each of these waters; they are all clear and agreeable to the taste, and mix well with soap.

Of eleven different springs derived from clay slate, which I tried in various parts of the province, including the six waters above enumerated, the specific gravity of nine was found to be 1·001; and the other two, from Paruree and Kumora (13 and 16), in the valley of Shore, were each 1·0014.

* This water was used exclusively by the sick, and by the servants attached to the hospital.

† Used generally by officers, servants, and other camp followers.

‡ Used by the 6th Company, 30th Regiment, during the residence of the regiment in Kemaon.

§ Used by the Light Company of the regiment, during the same period.

Effects of Reagents on these Waters.

a. They have no effect on the colour of litmus.

b. Lime-water affords no sensible precipitate when mixed with any of these waters.

c. Cold infusion of Brazil wood is changed from red to light blue by Paruree and Boorikote; but the other waters only render it bluish grey.

d. The natural colour of turmeric is unchanged by any of these waters.

e. Prussiate of potash occasions no alteration, with or without the addition of acid or alkali, in any of these waters; but tincture of galls occasions a slight precipitate in those of Paruree and Boorikote. This is, during the first six hours, light-coloured; it then gradually darkens, and in about forty-eight hours, it is found to be greenish brown.

f. Nitrate of barytes occasions no change in the appearance of any of these waters.

g. Oxalic acid discovers no precipitate in any of them.

h. Nitrate of silver occasions precipitates in these waters, which, with one exception, are more or less insoluble in nitric acid; the insoluble precipitates are in the following quantities, from three ounces of each water:—Large well, east of the lines at Lohooghat, $\frac{1}{3}$ of a grain; Boorikote spring, $\frac{1}{8}$ of a grain; well, north of the lines, Lohooghat Hospital spring, and spring between the hospital and the lines at that place, each $\frac{1}{10}$ of a grain. Paruree spring affords, with this test, slight precipitate of a brownish hue; but the whole is dissolved in nitric acid.

i. Acetate of lead affords copious precipitates from all these waters; and the following are the least soluble of them in acetic acid:—1, Boorikote; 2, Paruree: the others are entirely dissolved in distilled vinegar.

k. Alkaline carbonates afford only a slight separation of earthy matter from Paruree water.

Inferences.

1. From experiments *a* and *b*, in connexion with those of *c* and *d*, as well as from some of their physical properties, it is pretty evident that these waters contain little or no disengaged acid.

2. Experiments *c*, *h*, and *i*, are indicative of a minute proportion of a neutral salt, and a trace of muriatic acid is discovered by experiment *h*, in all but Paruree; in which, experiment *i* indicates a slight trace of sulphuric acid. This indication is not confirmed by experiment *f*; but acetate of lead being more powerful than nitrate of barytes in detecting sulphuric acid, the presence of a sulphate in this and Boorikote waters is scarcely to be doubted.

3. From experiment *e*, there cannot exist a sensible portion of any of the metals in these waters, unless the change that takes place in those of Boorikote and Paruree, under the application of the tincture of galls, be considered a proof of the presence of iron. It is, however, in confirmation of the presence of sulphuric acid, as indicated by experiment *i*: the ferruginous indication being here probably modified by sulphate of lime.

4. From experiments *g* and *k*, earthy matter is not contained in any of these waters, except in Paruree; and here the quantity detected by *k* must be very trifling.

The following solid extracts were derived from the evaporation of 13,000 grains of each of these waters, at a gentle sand heat:—

1. Paruree afforded 1 grain, which was constituted as follows:—	Alkaline and earthy carbonates, $\frac{1}{3}$; sulphates of lime, $\frac{2}{3}$ nearly; iron, a trace.
2. Boorikote afforded $\frac{3}{4}$ of a grain, which was constituted as follows:—	Earthy carbonates, $\frac{1}{3}$; sulphate of lime, $\frac{2}{3}$; iron, scarcely a trace.
3. Hospital spring at Lohooghat, 1 grain, composed of,	Sulphate of lime, $\frac{2}{3}$; muriate of soda, $\frac{1}{3}$.
4. Spring between the Hospital and the lines at Lohooghat, 1 gr.	Sulphate of lime, $\frac{1}{2}$; muriate of soda, $\frac{1}{2}$.
5. Spring east of the lines, Lohooghat, used by the 6th Company, 30th Regt. N. I., $\frac{3}{4}$ of a grain.	Muriate of magnesia, $\frac{1}{3}$; muriate of soda, $\frac{2}{3}$.
6. Spring north of the lines, Lohooghat, and used by the Light Company, 30th Regt., 1 grain.	Clay, $\frac{1}{3}$; sulphate of lime, $\frac{1}{3}$; muriate of soda, $\frac{1}{3}$.

It now appears, from comparison, that there is a difference in the nature and quantity of extraneous matters contained in the waters just noticed, which, though slight, yet is sufficient merely to distinguish them from the waters of limestone.

No doubt, my want of skill in the performance of chemical analysis, as well as the want of adequate apparatus, tended to render success much less complete than it might have been.

Appendix to the Analysis.

In connexion with the foregoing inquiries, I instituted a few preliminary comparisons between such distilled water as I could procure, the water of melted snow, and rain-water.

As the result of these experiments, in regard to rain-water, is different from what has been observed by others, it may not be devoid of interest to describe the process adopted, in order that the nature, or at least the value, of the difference in question may be rightly estimated.

The snow-water was procured in glazed earthen vessels, soon after a heavy fall that took place about the 20th December, 1833; after having been melted, the water was placed in bottles, carefully corked.

The rain-water was collected in glazed vessels, with every care, on the 7th July, 1834, a day on which nine inches of rain fell at Lohooghat; during the previous day, four, and the succeeding day, four-and-a-half inches of rain fell at the same place: so that any impurities this water contained may be supposed to have been derived from the general qualities of the atmosphere. This water was left a month or six weeks in loosely-corked bottles.

A crystal, weighing 3 ozs. 6 drs. 4 grs. in air, at 72° Fahr., was weighed in each of these waters, at the same temperature, and found to be 2 ozs. 2 drs. 41 grs. in the rain and distilled waters, and half a grain heavier in the snow-water. Each of these waters gave a perceptibly blue tinge to the infusion of Brazil wood; the snow-water, however, more obscure in its effects than either of the others, which, together with its lighter specific gravity, induced the belief of its being the purest of the three.

In order to ascertain the cause of the effect of the rain-water on the delicate test of Brazil wood, and recollecting that Bergman had discovered rain-water to contain muriate of lime—that Morveau had discovered sulphate of lime in it, while in England it has been usually found to contain carbonate of lime—I was anxious to ascertain the nature of the impurity of this fluid in a part of the world so remote from the places where the other trials had been made, as well as to know how far rain-water might answer as a substitute for distilled water in such inquiries as those in which I was engaged.

1. 13,000 grains, by measure, of rain-water were evaporated spontaneously to 1,000, in a broad porcelain dish, when a very slight deposit was found to have taken place, by the rough sensation it occasioned to the end of the finger on touching the bottom of the vessel

The supernatant fluid was decanted and evaporated to dryness by a gentle sand heat, when half a grain of a grey precipitate was afforded.

2. This precipitate was at first partly soluble in cold water, slightly deliquescent, and insoluble in distilled vinegar. On standing for a time, it became dry, and assumed the property of effervescing and dissolving quickly in acids.

3. After exposure to the blue flame of the blow-pipe, it loses the property of effervescing with acids, nor does it acquire the property of lime; for it retains its solid and compact form if immersed in water: if boiled, its size is increased rather than diminished.

4. About a tenth of a grain was dissolved in sulphuric acid, when, after standing, a minute particle of sulphate of lime was deposited, nearly equal in quantity to what a thirtieth part of a grain of carbonate of lime would afford. The clear acid solution was then rendered turbid by the addition of carbonate of potash.

Second Trials.

1. The dish in which the first part of the evaporation was conducted, and to which a slight precipitate adhered, was washed with half an ounce of the same water: the whole was filtered, when scarcely any solid matter was collected. The clear solution was first tested for sulphuric acid, by a solution of the nitrate of barytes, when no precipitate or change took place. Alkaline carbonates, if present, were then saturated with nitric acid, and nitrate of silver added, when an instantaneous precipitate was formed, which thus detected muriatic acid.

2. Another solution was now made by dissolving, by

means of nitric acid, a small portion of precipitate No. 1 of first series of experiments in a few drachms of the same water, when oxalic acid barely afforded a slight indication of lime; but which, from the experiments already related, may, as far as can be determined from experiments on such a small scale, be considered as carbonate of magnesia.

If the muriatic acid and lime, both of which substances were detected in the above experiments, were alone the only contents of this water, they would be melted on the slightest application of heat; but on mixing equal parts of carbonate of magnesia and muriate of lime, a compound is formed which is infusible before the greatest heat of the blow-pipe, and which presents analogous characters to those of the extract from the rain-water. Thus the presence of magnesia is probably indicated in rain-water.

ON THE CONNEXION BETWEEN GOITRE AND CRETINISM—THEIR NATURE AND CAUSES.

60. From goitre, as it appears in Kemaon, in its more distinct form, as well as in conjunction with cretinism, there are many reasons for believing that both complaints are intimately connected with each other; if not identically the same, they are mere modifications or different degrees of intensity of the same causes.

It may be remarked, that in those little hamlets where goitre prevails to a certain extent, the people are characterised by a want of enterprise and bodily vigour, as compared with their immediate neighbours, who are exempt from the disease. The distinction in this respect increases, not always, but in general, with the extent and severity of goitre, until at length both mind and body become so enfeebled and deformed, that the cretin is scarcely to be recognised as belonging to the human species.

61. Mr. Bramley, in his excellent account of the goitre in Nepaul (an adjoining kingdom), remarks, in a note, that he never saw cretinism, or anything approaching to it, in that country. If, instead of being attached to the court of a native state, Mr. Bramley had been so situated that he could have passed from the capital into the interior, and there pur-

sued his inquiries in the huts of the scattered population, he would probably have seen cause to express a different opinion. Nay, if I had been guided by information derived from old residents in Kemaon, rather than by my own labours, this paper would have contained a similar assertion; and as to information received from the common natives of India in particular on anything relating to statistics, it is not only not worthy of credit in a scientific point of view, but had better, in such investigations, be dispensed with altogether.*

Most of the conflicting opinions relating to this disease have arisen from authors and travellers resting their facts on no better foundation than that of the mere statements they derive from others, and thus but too often make popular error the basis of general conclusions.† I was assured, by persons for whose opinions I have great respect, that no such beings as cretins existed in Kemaon; yet, in the course of my inquiries (in which I made it a rule to take nothing for granted that I did not see and prove by the evidence of my own senses), I discovered whole villages of these unfortunate people.

62. In Goseragong the people are generally affected with goitre, yet there are no cretins among them. The same may be said of Deota; but in the villages of Salmora, Oliel, Goraght, Tomilly, and Ager, which contain 138 inhabitants, 76 have goitre, and 42 are cretins; while there is not one of the latter class to be found in any of those villages that are exempt from goitre.

Hence it appears, that in a population in which goitre prevails to the extent of rather more than 50 per cent., 30 per cent. are cretins; while, in the remaining portion of the people, amongst whom goitre does not extend to above 9 per cent.,

* I make these remarks generally, in order that we may avoid, as much as possible, a very common source of error.

† Prom Pileabit to the confines of Rohilcund and Hurdwar, is stated, by Mr. Bramley, on the authority of another author of great respectability, to be a link in the chain of affected districts in the plains of Hindostan, extending from 27° to 30° N. lat.; yet I have traversed the Tarai in this direction, from Pileabit to Burmdeo Pass, and from thence to the vicinity of Rudirpur, crossing from thence to Moradabad, without having met with a single case of goitre, although I made it a point to visit every village of the Tarai during a march of at least 150 miles. Fairs are, however, held in the Tarai during the cold season, at which the inhabitants of the mountains and other places attend; and in this way the mistake may have arisen.—See *Cal. Med. Trans.*, vol. vi., p. 182.

we have no cretins. Thus far, it is clear that the two diseases are connected with each other, not merely endemically, but they are combined and blended together in the same individuals.

63. In describing the disease (Par. 49), it is stated that children are exempt from it until the age of three years. This is also in conformity with the observation of Mr. Bramley; nor has any authenticated instance occurred of congenital goitre, although an analogous condition of the bronchial gland of some of the lower animals is congenital, a circumstance which of itself forms a distinction between the nature of the disease of animals and the goitre of the human subject, that we cannot possibly overlook.

64. Delicate (apparently), ill-fed, and neglected children, in certain villages, become affected by the disease in the course of a year or two after they are taken from the breast. It is usual for them to have long matted hair, large joints, tumid abdomens, and slender limbs.

The tumour on the neck makes greater or less progress for a time, but usually becomes interrupted before it attains a larger size than that of an orange, and the general health now rapidly improves.* In other cases, numerous bronchial glands are simultaneously attacked, and the augmentation of the tumours, which soon unite, suffers no abatement; while the general development of both mind and body is for the time suspended, or the materials of the latter are rather directed to the formation of irregular accumulations, generally on the neck, than to the uniform increase of the body. Nor does this morbid action, or *error loci*, suffer any interruption until the subject has attained the adult age.

65. With respect to the first of these cases (Par. 64), the interruption to the growth of the tumour does not take place sometimes until it has reached its full size. The necessary period for this varies from 10 to 30 years; and often the tumour continues slowly to increase during the life of the

* Alibert divides the tumours into simple and compound, according to the nature of their contents. Compound bronchocele illustrates these views—Alibert having found such tumours to contain calcareous, sarcomatous, and fatty matters, as well as other heterogenous contents, such as hair, &c. Human dissections not being tolerated in Kemaon, I can offer no remark on the pathological character of these tumours.

patient, but so insensibly, that, at an advanced age, it is frequently found of an inconsiderable size. In such cases, the general health continues good; and hence, even in villages, where the exciting cause may be supposed from the number affected to be very intense, we often find strong, robust, and otherwise healthy adults with goitre of every size and shape—a circumstance which has erroneously induced some to believe, that the disease is merely local; and as these are the sort of cases that usually occur to common observation, the error in question is by this means rendered the more general.

66. The second variety of the disease (Par. 64), or that which occasions the peculiar condition called cretinism,* is distinguished from the last described by a greater intensity of all the symptoms. The patient is invariably, and indeed necessarily, seized, during the first stage of life, *i. e.*, before the age of five years; and the disease continues without interruption throughout adolescence. During this time, there is a depraved action of the absorbent system, shown in the monstrous development of certain organs; while the natural growth of others is proportionately suspended. Hence, the limbs are short and crooked, the spine distorted, the head often of enor-

* Foderé, and others, ascribe the weakness of the mental energy of cretins to the state of the thyroid gland—an opinion which Mr. A. Burns, as well as Mr. Cooper (*Surg. Dic.*), very properly suspect to be without foundation, from the fact of cretins having been seen without much enlargement of the thyroid gland. These eminent surgeons were not, however, justified, on this account, in considering the connexion between cretinism and goitre as merely accidental; as if mental imbecility were an essential symptom of cretinism.

The cretins in Kemaon are characterised by *general deformity of the body; but especially* of "the head and neck; countenance vacant, and stupid; mental faculties feeble, or *sometimes* idiotic"; sensibility obtuse; mostly with enlargement of the thyroid gland.

This description, with the exception of the words in italics, is from Dr. Good's *Nosology.* The deformity of the head, a symptom of the general disorder, may of itself give rise to "the mental faculties feeble," and "sensibility obtuse," as it is only in those who are thus deformed that these symptoms are very apparent.

It must, however, be remarked, that the "countenance vacant and stupid," in Dr. Good's characters of cretinism, may often be only the false effect of disproportionate features. In proof of which, I have only to mention that, although the cretins of the village of Ager have these characters in an extreme degree, yet they perform the practical duties of working a copper mine in their vicinity—an occupation in which they display at least some mechanical skill.

mous size, the features bulky and idiotic, and glandular swellings are common on various parts of the body, but seldom entirely absent on the neck, where the first signs of the disease are displayed in the enlargement of the bronchial glands.* This malformation usually continues until the end, or, if life is spared, to the adult age, when it is not further extended.

Such physical derangement, affecting three-fourths of a whole community, is calculated to impair, in a moral point of view, their intellectual faculties; there is still, however, reason to fear that, in the majority of cases, both mental and corporeal functions suffer alike. Yet the cretins do not, in general, equal the imbecility of natural or born idiots; but on the contrary, they retain at least some glimmering of reason acquired in childhood, and which enables them to perform many useful offices which their unfortunate condition and numbers must call so frequently into requisition.

67. From the above description of the disorder, it must be evident that goitre and cretinism are but varieties of the same disorder, and that the proximate cause of both is an *error loci*, or derangement of the functions of the absorbent system.

On the Remote Cause.

68. In conformity with the custom of pathologists, the remote cause may be divided into predisposing and exciting causes. In the present instance, indeed, this division is indispensable.

Predisposing Cause. "There are many reasons to induce us (says Dr. Robertson) to regard goitre as a particular variety of scrofula; in this country (England) it is only seen in highly scrofulous constitutions." Now, although I cannot venture to say that goitre is only seen in Kemaon in highly scrofulous constitutions—yet I must bear testimony to the accuracy of the remark to a certain extent; and beyond this, what is stated

* Cretins are sometimes seen without any great enlargement of the bronchial glands; but such cases are rare, and they are generally otherwise much deformed, so as not to be mistaken for congenital idiots. Tumours on the elbows, knees, and other joints, as well as along the course of the lymphatics, are common with the cretins of Ager (Par. 43), and other villages in Kemaon.

in para. 65 will explain the cause of numerous healthy, or at least stout healthy-looking, persons being seen with goitre. The opinion of Dr. Robertson has often been suggested without assigning adequate reasons, and as often opposed on still more inadequate grounds; while the great bulk of those to whom arguments on both sides were addressed were unable to decide, for want of practical acquaintance with the points at issue. I shall here transcribe from Cooper's Surgical Dictionary the distinctions stated to exist between goitre and scrofula, as enumerated by Dr. Pastiglione; and to save repetition, I shall take the liberty to refer the reader to the articles in this section, by comparison with which, each of the supposed distinctions will be found to give way.

1. "Scrofula is a disease of the general system, but bronchocele is merely local."

This distinction is removed by what is stated in para. 65, where the error is accounted for, and explained: *see* also paras. 64, 66, and 68.

2. "Bronchocele begins at a later age than scrofula, and does not, like the latter, spontaneously disappear."

This distinction is completely removed by what is stated in paras. 64, 65, and 66.

3. "Scrofulous glands often suppurate; bronchocele rarely undergoes this change." This is the only real distinction.

The only other distinctions between scrofula and goitre, pointed out by Dr. Pastiglione, are very trifling; and the whole of them merely refer to the difference between the simple form of goitre and scrofula: but if the connexion of the former with cretinism be granted, as I believe it must, the difficulty of longer drawing any sound distinction between these diseases, except as varieties, is much increased.

Exciting Cause. This cause has been traced, in the foregoing inquiry, to certain strata of the earth, under circumstances calculated pointedly to suggest, that water is the medium by which it is conveyed to the bodies of men; although the analysis of such water has proved insufficient to detect any ingredient to which we can directly ascribe their effects (paras. 55, 58), except lime.

Finally, that having thus far traced the source of the endemic, we have reached, in regard to the exciting cause of goitre, the utmost limit of our knowledge: but whether there be any other strata capable of yielding this peculiar contagion than those we have described, and whether the water is the only medium by which it is conveyed, are points which still remain to be determined.

GOITRE AND CRETINISM IN THE PLAINS OF BENGAL.

At the time the paper on goitre and cretinism appeared some years since in a work printed in India, I felt then little inclined to give the results wider circulation, in hopes of adding to their interest by more extended inquiries in other provinces where goitre was known to exist, more especially in the plains.

The care and pains with which these results were worked out, left little doubt but that more extended inquiries would eventually confirm their accuracy.

I am far, however, from supposing that we have not still much to learn on this subject; my only object has been to find out if possible, the right path to follow in the inquiry, and then to leave the further investigation of the subject to others.

The observations of Professor Kölliker, referred to in the Preface, are likely to produce a change in our views relative to the proximate cause of goitre and cretinism, more in harmony with the general tenor of these inquiries.

ON THE INFLUENCE OF SOILS.

FIVE years after the observations above detailed were printed in Calcutta, Mr., now the Rev. Professor Liston of Edinburgh, then in India, communicated to a journal with which I was connected, a paper on the distribution of soils in the Goruckpore district, which is of so much consequence, and bears so strongly on the present subject, that I may be pardoned for taking the liberty of quoting the following.

Those lines of Professor Liston's paper having peculiar interest, as affording the first and only suggestion ever made, up to this time, touching the existence of cretinism in the plains of Bengal, I have placed in italics.

"The Goruckpore district is bounded on the west and south by the river Gogra; and three-fourths, or it were perhaps more correct to say seven-eighths, of the Zillah are watered by that great river, and by the Raptee and other tributaries which fall into it from the north. On the east, the district is bounded by the Great Gunduk, and the remaining portion is intersected and watered by tributaries and branches of that noble stream. But what seems rather a remarkable circumstance, the lands under the rule, so to speak, of the Gogra and its subordinates, are in general all of a description of soil called *Bangar*, that is, of a dry silicious nature, and requiring irrigation for the production of rubbee or winter crops, whilst the lands bordering on the Gunduk and its branches or feeders are what is termed provincially *Bhat*, a soil retentive of moisture, producing cold weather crops without artificial watering, and a very notable portion of it being calcareous matter.

"The Little Gunduk may be considered as the boundary of the Bhat or calcareous deposit to the west, as I believe I am correct in stating that this description of soil does not obtrude on the left side of the river in its whole course, beyond what may have been an old bed of the river, or where back-water from it may have been stagnant during wet seasons. The Little Gunduk falls into the Gogra, and therefore, agreeably to the above general observation, the lands upon it should partake of the nature of those on that river; though the main portion of its stream comes from the lower range of the Nepaul Hills, and is thus so far independent of the Great Gunduk; it also draws a part of its supply from the latter river in the rains, and the lands bordering on it are of a mixed nature; in fact, we may consider the characteristic of the soil bordering on the Little Gunduk as calcareous when nearest the Great Gunduk, but silicious as it approaches the Gogra.

"It strikes me that the portions of the country where silicious soils prevail, abound more in kunkur beds or ridges than where the calcareous soils are met with, however they may only be concealed in the latter case; but admitting the matter to be as now stated, a query arises as to whether the lime incorporated with the soil in the Bhat division of the country has in the Bangar portions been drawn down, while the country was in a state of submersion, to form the kunkur beds which are there met with? Another question occurs—how came the kunkur bed to be formed at all? Were the nodules in which it is generally found, formed on shells as nuclei? In looking at some of the kunkur knolls in this district, one would readily fancy they originated in this way, the groundwork of them being from the materials of shell fish, &c.

"As this Bhat deposit is confined to the country of the Great Gunduk and its branches, being bounded on the east, as I am well informed, by the Bhogmuttee in Tirhoot, and extending down the Little Gunduk in that portion of the country to nearly opposite to Monghyr (and let it be observed, there are two Little Gunduks, one to the east and another to the west of the Great Gunduk), the appearances now mentioned might be accounted for by supposing a series of high floods in the Great Gunduk, which caused it to fill and overflow its own bed, and that of other streams to which it may have had access, occasioned, say, by the bursting of barriers in the upper portion of its course (and it drains a great portion of the Nepaul territory), behind which masses of loose calcareous matter may have accumulated, and which the water as it escaped swept along with it. But, though such hypothesis would explain the fact of a pretty uniform sheet of calcareous matter covering the region now spoken of, it will not clear up all the appearances; for there are alternations of soil, as of Bangar over Bhat, to be met with in some places, which show that various changes in the condition of the tract have taken place.

"North of Selempoor, and on the left bank of the Little Gunduk, small eminences abound; these have generally been selected as sites for villages, and are Bangar, or silicious earth, while the lower lands are Bhat. But, I am told, in digging wells on these elevations, after the Bangar strata are penetrated, Bhat earth is met with. Did, then, a uniform sheet of silicious soil at one time cover this section of the country, which has subsequently in great part been washed away? If so, how come these knolls to form exceptions? Or, if not, how came they in their present places?

"The Bangar countries, so far as I have observed, undulate more than the calcareous districts, and the water shows a disposition to collect into shallow lakes. The water in the Bhat countries rather affects to form itself into streams, hollowing out for itself in the rains, canal-like beds, through which water that would otherwise be stagnant is drained off into the larger rivers. The Bangar country is more congenial to the human constitution, and, I may add, to horses and dogs also, than Bhat lands are; *the inhabitants are better grown, stronger, fairer, handsomer, and the children more playful in the former, than in the latter.* In the calcareous damp lands, goitre prevails, idiots are common, and, in general, the people are feeble in mind and in body.

"The following analyses of some soils will serve to give a more definite idea regarding the land in different parts of the country than is to be got from more general observations:—

"*Analysis of* BANGAR SOIL *from Selempoor (specimen long gathered)*—

Water of absorption	1·75
Taken off by filtering solution in distilled water ..	2·75
Carbonate of lime	0·2
Carbonate of magnesia	0·25
Oxide of iron	2·2
Alumina	9·75
Silex	80·
	96·9
Loss	3·1
	100·

"This specimen was gathered from a patch of uncultivated ground, but which would grow good poppy or sugar cane, &c.

"*Analysis of a stiffer soil,* MUTTEEAR BANGAR, *from a lower site in the same neighbourhood (specimen had been exposed to rather damp weather before being manipulated)*—

Water of absorption	5·3
Taken off by solution in distilled water	3·5
Carbonate of lime	1·
Carbonate of magnesia	0·5
Oxide of iron	2·75
Alumina	13·6
Silex	71·25
	97·9
Loss	2·1
	100·

"*Analysis of* BHAT SOIL, *from Perruna (specimen long gathered)*—

Water of absorption	3·25
Taken off by filtering solution in distilled water ..	2·5
Carbonate of lime	25·3
Carbonate of magnesia	2·
Alumina	7·
Iron	0·25
Silex	57·75
	98·05
Loss	1·95
	100·

"Soil adapted for growing Falginee indigo (*i. e.*, sown in March, and not irrigated) as in the Tirhoot factories, which are generally on this kind of land, and confined to the calcareous region upon the Great Gunduk.

"It may also be worth stating, that matter deposited in the reservoir of an indigo factory supplied with water from the Gogra, when treated with dilute muriatic acid (2 water, and 1 acid), showed a loss of 7·5 per cent.; so that river seems to bring down much more lime than the soil in its neighbourhood —*i. e.*, six or eight miles from it—contains. Again, matter deposited in a reservoir supplied from the Little Gunduk at Selempoor, when similarly treated to the above, showed a loss of 27·5, and thus seems to carry with it mud as rich in lime as the soil about Perowne."

Attention was called at the time to the great interest of the above paper, in hopes of obtaining further information from the local authorities, and others on the spot, but without effect.

After a lapse of some years, finding the subject neglected and forgotten, chance afforded me an opportunity, while serving with the Goorkah force in the suppression of the late mutiny, of observing the prevalence of goitre and cretinism in the districts of Goruckpore and Champarun.

The occasion was unfavourable for the collection of such full, and complete information on the subject, as could be desired.

As goitre is not a disease the existence of which demands very close observation to detect, so our knowledge of its prevalence in the districts from Rungpore to Goruckpore, is almost coeval with that of the districts in question.

These districts extend along the borders of the Tarai, and this last extends along the foot of the

Nepaul mountains; and so, from an association of ideas, the goitre has been here generally supposed to bear some immediate local connexion with these mountains.

But such is not the case, for it is found in open cultivated districts of the plains sixty miles from the mountains. In other places it may extend much closer to them, but never so as to afford any reason to suppose that it can be due to their influence, merely as hills.

As our European troops are now stationed in some of these districts, the foregoing facts connected with this important subject may prove of some additional interest and use, in suggesting the necessary steps for guarding against exposure to this disease, as well as cretinism. The description given of goitre, para. 49, and of cretinism, paras. 60 to 67, as these diseases appear in Kemaon, apply equally to them as they occur in the plains of Bengal.

From all I saw and could gather on inquiry, I should think, upon a moderate calculation, the number of children who annually, if not born cretins, yet fall into that state in their infancy in these districts, must be great.

In one locality alone, during a halt only of a few hours, Mr. K. M. Nicholson of Bubnowlee, in the Goruckpore district, collected numerous children—*bowkuts*, or cretins—from an adjoining village, for my inspection, as examples of the condition of the people in large tracts of country in his neighbourhood.

The following memorandum on the subject was drawn up on the occasion from the general result of observation and inquiries in these districts:—

Goitre, or *geega*, as it is called, is very prevalent

where the soil is white, damp, and sandy; such lands being called *bhat*, as not requiring irrigation for the dry weather crops. It is unknown in Bangar lands, where the soil is reddish, sandy, and dry, so as to require irrigation in the dry weather. There is, however, no remarkable difference of level in the two kinds of land generally.

All the villages on the *bhat* lands are not equally affected; the inhabitants of some being quite exempt from these forms of disease, while the inhabitants of others are affected to the extent of 10 per cent. of their population; women being more subject to the disease than men.

Ten per cent. of the children in villages where goitre prevails are *bowkuts* or idiots. Many are deaf and dumb; their limbs weak and feeble; their features large, and without expression; the eyes dull and vacant. They seldom live to attain an adult age.

Cretin children in these districts are generally, but not always, born of goitred parents. Sometimes one parent has goitre; more rarely the parents are free from that disease. Cretinism begins to show itself at a very early age; the goitre comes on later. Both forms of disease sometimes go together in the same individual; but there are many instances of cretins without goitre, and numerous cases of goitre in otherwise strong, and to all appearance healthy persons.

Dogs and cats are often affected in these districts with goitre. Certain wells are known to produce goitre, although the water presents no difference in appearance from that of other wells, nor can any be discovered in taste.

Such are the general circumstances under which goitre and cretinism appear in the plains of Bengal; they will be found to differ little from the form in which these maladies appear in the mountain province of Kemaon.

In treating of the geological structure and composition of the soils composing the plains of Bengal and the N.W. Provinces generally, it was observed, with reference to Goruckpore and the other districts in which goitre and cretinism prevail, that the calcareous clay is far more lightly covered than in the Doab, and forms considerable tracts of the surface soil, under the term of *bhat* lands. This observation was made without the slightest regard to the influence of the soil on the human constitution, but was merely stated in the course of inquiry, and amply confirms the observations of Prof. Liston.

The difference, therefore, between *bhat* and *bangar* lands is due to the damp clay bed being in the one case lightly covered with dry soil, and the other deeply covered with dry sandy deposits. But damp soils in other parts of the plains do not produce that derangement of the glandular and osseous system on which goitre and cretinism depend, because these maladies are unknown in the damp clayey districts of Bengal, Assam, Pegu, and other parts of India, while they are prevalent in certain dry rocky soils in Kemaon, where the waters employed by the inhabitants are derived from beds chemically identical with the calcareous clay bed composing the *bhat* lands of Goruckpore.

Thus we could have no better confirmation of the

influence of soil on the human constitution, than to see the same agency operating alike in the mountains and the plains, through the influence, we may almost venture to conclude, of water derived from springs in certain calcareous rocks and soils.

In an artificial or highly civilised state of society, there may, indeed, be other causes exercising a more direct influence on health than such as depend upon soil. But in communities reduced to a more simple state of life, we may perhaps, be less surprised to find the human constitution more susceptible to such influence. It is in such conditions of society that we may eventually hope for the most successful results, arising from inquiries like these.

Remittent Fever.—It has been remarked, while passing from Midnapore to Sumbulpore, throughout the tract of country noted for. the fatal character of its fevers, that the soil consists of red ferruginous clay.* This is the laterite so generally distributed, more especially in low hilly tracts, where it occupies the intermediate levels between the flat alluvial plains and the rocky sides of the mountains.

It generally presents a broken, hard, rough surface when much inclined and sloping, or a smooth, greasy kind of surface when less so. In wet weather, and long after rain has subsided, it continues soft and muddy. The surface is bare when inclined; the only vegetation being a few bushes thinly scattered in drier situations, but collected into a more dense brushwood in the lower undulations. In flat, low, wet,

* Martin's *Influence of Tropical Climates*, p. 14.

uncultivated places it is always covered with a coarse grass.

It is not so much the mineral characters, as the physical properties of this clay that we should regard with suspicion, when we find it composing low, and ill-drained places, because we find its influence the same in the Nerbudda, and N. W. Provinces, where the calcareous element frequently prevails over the ferruginous in its composition.

There, the chemical components of the clay are very different from what they are found to be at the foot of the Rajmahal hills, and everywhere to the south of that position, but its physical properties and its influence with regard to remittent fever are the same.

The character in common which the remittents of these laterite districts present, and by which they are distinguished from the corresponding type of fever of the Tarai and the Sunderbunds; is, that while the fevers of the laterite districts prevail during the hot weather as early as March, those of the districts composed of light sandy soil do not set in until after the rainy season has commenced, being above two months later.

Again: the jungles of laterite districts do not become safe until the month of January—that is, three months after the rains have subsided—while those of light soils of the flat plains become safe in November, within a month or six weeks of the breaking up of the rains.

These, I am aware, may seem very common-place observations, but in a practical point of view they are not the less important.

The most opposite opinions frequently prevail as to the season when you may with safety enter, or prolong your stay in the jungles. One person will declare the hot season to be the safest time; another, that it is the worst, or most dangerous. Both are right; but many a valuable life has been lost for want of proper means of discrimination between such conflicting opinions, which due attention to the nature of soils will afford.

In the low hilly tract of Moispore, 40 miles west of Jungypore, consisting of undulating broken ground situated at the foot of the Rajmahal hills, the jungle becomes very unhealthy in March.

The soil consists of laterite resting on trap rock, which protrudes at intervals through the laterite, forming small rugged hills. On the higher smooth rounded undulations, the surface presented a singular hollow sound under the feet, which was so remarkable as to attract the notice of our native attendants, who could not resist the inducement of dancing, as they passed over these places—amused with the sound of the feet, which was something like that of a drum. Water is scarce, but occasionally a small stagnant pool may be observed in hollow situations.

The elevation of the district varies from about 80 to 150 feet above the plains of the Ganges, situated at a distance of 30 miles, to which the country sloped with a south-east aspect.

A party consisting of four Europeans and about fifty natives, arrived in this locality in good health on the 6th of March. As they had been chiefly all out during the whole of the preceding hot season in the upper valley of the Sone, they felt a confidence in their own experience, which it was impossible to dispute, that the hot season is the safest time to be in the jungle.

On the 11th of March remittent fever appeared; on the 12th, half the party were ill from the same cause; and, on the evening of the 13th, there were scarcely a sufficient number remaining to assist the sick, who had to drag each other out of the district during the night, by which timely retreat most were saved.

The Moispore district is formed of a continuation of the same red soil as that composing the hilly tracts referred to by Dr. Martin, between Midnapore and Sumbulpore, as noted for the fatal character of its fevers.

Both tracts of country are situated in the same division of climate—namely, that of Beerbhoom, &c., page 56—in which the annual fall of rain is not quite 29 inches. At Ranygunge, an intermediate point between these two bad districts, and in the same local climate, the continuity of the laterite is broken by sandstone, which forms the surface, covered with light sandy soil. Here we have a European station which is not remarkable for endemic remittents, thus proving that there is no inherent objection to low hilly tracts, apart from the nature of the soil.

Were the various tracts of country subject to these hot season remittent fevers to be traced, it would only be to repeat what has been said regarding the distribution of the laterite clay throughout the low hilly districts which border the plains and valleys of Bengal and the N. W. Provinces.

On the other hand, Hidgelee may be taken as an

apposite example of those districts in which the hot weather is the safest time. It lies on the western side of the entrance to the Hoogly river.

Taking Balasore as presenting the mean annual temperature of the neighbourhood, it is 3°·7 Fahr. higher than that of Agra, and 11°·4 Fahr. higher than the mean temperature of Chittagong in nearly the same latitude, during April and the next succeeding five months.

Hidgelee, therefore, combines the influence of a low sea coast with the highest temperature of the plains. It is enclosed within embankments, to prevent irruptions of the sea on the one side and of rivers on the other; yet, from the light, porous character of the soil, the waters dry up rapidly after the rains terminate, and it continues free from endemic remittents from November until the first fall of rain in June, when they suddenly appear in a virulent form.

The same observations apply equally to the fevers of the Tarai, along the foot of the Nepaul mountains, far removed from the sea-board.

I have already alluded to the fact of the Goorkah army having crossed this tract, on their return to Nepaul in the beginning of June 1858, without having suffered, although these mountaineers are as subject to the fever of the Tarai — or the *Oul*, as they call it—as Europeans. The only anxiety of the Goorkah officers on that occasion was, that they should get their men across before the rains commenced. The force began to cross on the 1st of June, and on the 6th the last portion had reached the mountains in safety.

This tract is composed entirely of sedimentary deposits, which soon dry up after the rains are over; and, like the whole of the low flat tracts composed of estuary and river deposits, whether on the sea-board or along the plains and rivers, become safe and free from endemic remittents after the middle of November.

Comparing the remittent fever of tracts composed of laterite with that of the sedimentary plains, we perceive at once the direct influence of soils upon malaria; in the laterite soil high temperature being alone sufficient to produce the endemic effect, but in the sedimentary soil that condition requires to be conjoined with a fall of rain.

The foregoing observations apply only to the influence of soil in connection with jungle fever. With regard to the fevers which European troops are subject to in cantonments of the plains, they depend partly, no doubt, on the local malaria, yet they are so greatly modified by other causes—such as affect the sanitary condition of the troops generally—as clothing, diet, drink, discipline, and the like, that they do not come within the scope of these inquiries.

Cases of fever arising from these complications and other similar causes are, however, of a more tractable and mild form. This is partly from the greater facilities at hand for prompt and energetic treatment at the commencement of the attack.

There is however, a tendency in all fevers, from whatever cause they may arise in the first instance, to assume the character of the endemics of the climate and place in which they occur. Hence the vast importance of every fact calculated to improve our knowledge on the subject of malaria. These fevers of military cantonments in the plains frequently assume the remittent type, and are shown by the official returns to be by far the most fatal disease to which the European soldier is exposed in India.

But all low plains are not composed of light sedimentary soils. Large tracts are composed of laterite, often uncovered by a lighter soil, in the eastern parts of Bengal, about Sylhet, and from thence to Dacca and Jumalpore; in Pegu, from Rangoon to Shewygeen in the Sitang Valley, and Therawaddi and Henzada, in the Valley of the Irawaddi.

Such low tracts dry up slowly, and are the seat of intermittent fevers in the rains, and of remittents more or less at all seasons, but more especially at the close and for three months after the breaking up of the rain, as well as in the hot weather.

Dysentery is also prevalent during the rains in these tracts, while wounds or other injuries of the soft parts, however slight, have a tendency to ulcerate.

The native inhabitants of low laterite plains evince an instinctive objection to live on ground-floors. Their houses are consequently raised on posts. This practice prevails in Eastern Bengal, Upper Assam, and in Pegu. It would seem to have originated in a perception of the capillary attraction of the soil, in consequence of which, the surface is always wet. Thus the natives of these provinces display a just appreciation of the influence of soil, which becomes ingrafted even in their national character and customs. Silicious soils do not possess this property; so that the surface, however slightly raised above the influence of inundation, is comparatively dry. In Western Bengal, the practice of building on posts is, therefore, quite unknown, however the locality may be exposed to inundations—a little bank of earth, on which the house is made to stand, being considered a sufficient protection from damp.

Dracunculus.—Let us now turn to another disease, from the effects of which European troops serving in tropical climates also suffer severely, and we shall find how intimately the influence of soil is connected in another way with the cause, and how the sciences of zoology and geology both tend to aid the medical practitioner. We know the disease to be occasioned by the presence of an *entozoa* in the cellular texture of the body. There can be no doubt, from the weight of evidence on the subject, that it is of endemic origin, or confined to certain localities more or less circumscribed.* Nearly all writers on the disease, as it occurs both in the West and East Indies, coincide on this point, as well as in referring it to certain peculiarities of earth and soil. Dr. Duncan states, that the district in which he observed the disease, is composed of trap rocks, which is borne out by the greater prevalence of the disease in Bombay than in the Bengal Presidency.

The next point to be considered is the season at which the disease appears. All writers agree as to the period of attack being from May to September. Dr. Morehead, however, states that at Kirkee, where the disease was prevalent in 1832, that it set in as

* *See* papers by Dr. James Bird, Dr. R. Kennedy, Dr. G. Smytton, and Dr. C. Morehead, all of the Bombay Service, in the 1st and 6th volumes of *Transactions of the Medical and Physical Society of Calcutta.*

early as March, the admissions into the hospital of the 4th Dragoons being—April, 7 cases; May, 57; June, 64; July, 48; August, 26; September, 3.

Another fact of importance, connected with the history of this disease, was brought to notice by Dr. Scott,* of its having appeared among the troops at St. Thomas's Mount in June, one year after they had returned from an expedition to the banks of the Tumbudra. None were affected but those who accompanied the expedition in question, with which they were absent from the Mount from February to June.

Hence it would seem, that a certain period, although not necessarily amounting in all cases to a year, is necessary from the first arrival of the troops in an affected district, for the development of the disease.

We have next to consider the nature and habits of the dracunculus itself, the immediate cause of this disease.

Dr. Duncan and other writers coincide in the fact that, as the time arrives for producing the young, the dracunculus works itself out of its nidus by means of the awl-shaped sharp-pointed tail, which is always the part presented externally.†

A single individual produces myriads of young, enveloped in a bluish-white fluid, a single drop of which contains thousands.

In some experiments made on the subject, a few drops of the fluid were placed on a glass under a microscope, and twenty-four hours after, the mucus

* *Edinburgh Medical Journal*, 1831.

† *Trans. of the Medical and Physical Society of Calcutta*, vol. vii., pt. 2.

was found to have become perfectly dry on the glass; it was then moistened with tepid water, when the young were again found in motion.

The size of the young, not a line in length, is still further diminished by their habit of coiling themselves up. Dry and torpid in this state, they are no larger than the smallest mote.*

Dr. Duncan has remarked that the tail of the young is armed with a sting, by which they fix upon any object, and, if appropriate to their nature, they can work their way into it.

Dr. Duncan found the young to live at least for some time in a glass of water, how long is not stated.

According to my own observations, when immersed in a drop of water they seemed to live so long as their mucous envelope remained unbroken; but when that happened, and they became exposed to the water, they soon expired.

The young, although so minute as to require a microscope of some power to see them, are perfectly formed in all respects like the old, but display far more energy and life.

From the foregoing circumstances connected with the endemic nature of the disease, it is highly probable that the young entozoa discharged into the air during the hot season, may find an influence in trap rocks, favourable to their preservation in a torpid state. This may arise from the low conducting power, as regards heat, possessed by these rocks, their consequent high and uniform temperature in tropical climates,

* *Calcutta Journal of Natural History*, 1840, p. 359.

together with their softness to the touch, and the absence of hard sharp sand in their débris.

The season of the attack, and the time the disease lies dormant in the system after exposure to affected districts; the habit of discharging the young into the air; their power of preservation in a torpid state; their lightness and minute size, and the restoration of their vital energies on coming into contact with a warm humid body; their power of attaching themselves to it;—all tend to one view of the mode of attack and progress of this disease, namely, that the young are preserved in a torpid state in certain soils or rocks; and in the hot weather are carried, perhaps by the wind, or otherwise brought in contact with the body, into the surface of which they penetrate by means of the sting-like tail, in the manner described by Dr. Duncan; thus affording, as he remarks, an easy solution of the opinion, that attendants on patients, and dogs moving about them, get the disease.

Elephantiasis is another form of endemic disease upon which some light may in future be thrown by investigations of this nature, conducted in the districts to which it is confined.

Conclusion.—Enough, however, has been said, to point out the deep importance or absolute necessity of extending the basis of all inquiries relative to the nature and causes of endemic diseases in tropical climates, so as to take in a wider range of facts than has usually been considered necessary in such researches. It is with a view of promoting this object that the following table of the stations in Bengal and the N.W. Provinces, as well as Pegu, arranged according to the rocks and soil on which they stand, is submitted; subject, of course, to correction in some cases, and extension in others, so as to include stations and facts which may be here overlooked.

The numerous able and zealous medical officers of the other presidencies will readily, if called upon, supply corresponding tables of the stations of Madras, Bombay, and the Punjab. Thus provided with such information, the words "Laterite covered with sandy soil," or, "Yellow clay resting on granite," as the case may be, at the head of Hospital and Sick Reports, or other Medical Returns, would give a value to such documents which they do not now possess, and might lead to the most important results in the extension of medical science.

Table showing the rocks and soils on which the undermentioned stations stand in Bengal and the N.W. Provinces, as well as in Burmah.

Laterite covered by light sandy soil.	*Gneiss covered by laterite and sand.*
Jungypore.	Hazareebagh.
Burdwan.	Gyah.
Midnapore.	Monghyr.
Jumalpore.	Gowahatti.
Mymensing.	Meeaday.
Thyetmyew.	
ToungHoo.	*Trap covered by clay and dark soil.*
Moulmein.	
Rangoon.	Neemuch.
Pegu.	Mhow.

Kunkur covered by deep deposits of fine sandy soil.

Meerut.
Umballa.
Agra.
Delhi.
Allahabad.
Benares.
Ghazipore.
Buxar.
Dinapore.
Lucknow.
Fyzabad.
Sultanpore (Oude).
Bareilly.
Goruckpore.
Mozufferpore.
Segowlee.

Delta deposits.

Berhampore.
Chinsura.
Barrackpore.

Delta deposits (cont[d]).

Dum Dum.
Calcutta.
Sylhet.
Dacca.

Sandstone covered by laterite and fine sand.

Chunar.
Nowgong.
Saugur.

Clay slate.

Almorah.
Darjeeling.
Lohooghat.
Petorahgurh.

Argillaceous clay covered by fine sand.

Prome.

Sandstone.

Cherraponjee.

GENERAL REMARKS.

Stations in the Plains.—We have no reliable data to show, on a sufficiently extended basis, the relative healthiness of one station compared with another; besides which, such results depend on so many collateral causes connected with the condition of the troops at the time of their arrival at any given station, their previous exposure, their discipline, barracks, temperance, and the like, that a station which, at one time, bears a favourable character, may, at another, be regarded as unfavourable, from circumstances quite unconnected with its climate. It is scarcely probable, however, that a train of accidental circumstances, such as those alluded to, should extend through a series of years, to keep up a certain character of any given place, without the local climate being in someway concerned in the result. In this point of view, the annexed table, which exhibits some of the results brought forward by Sir A. Tulloch, from the War Office returns, are sufficiently striking to demand the most serious attention.

It would seem to be impossible that a difference of 32·26 per thousand in the relative mortality of troops stationed at two points on the same plain, as Cawnpore and Meerut, extending steadily over a long series of years, can exist without some obvious and distinct cause, capable of being pointed out when properly inquired into. It is true, the cause when known, might be difficult to remedy; but still it ought to be ascertained, were it only as a guide for the future selection of other stations.

	Period of Observation.	Ratio of Deaths to 1,000.
Calcutta	24	65·92
Chinsurah	9	56·64
Berhampore	14	63·93
Bhagulpore	4	87·96
Dinapore	12	78·89
Ghazipore	5	91·94
Cawnpore	20	66·13
Agra	12	47·26
Meerut	33	35·87
Kurnaul	11	46·30
Hill Stations.		
Hazareebagh	4	47·07
Kussowlie	7	49·01
Dugshai	3	26·36
Subathoo	3	31·02

The intermediate station of Agra, holds also an intermediate place in regard to the ratio of deaths, while the neighbouring station of Kurnaul assimilates with the more favourable character of Meerut, thus rather favouring the probable existence of some very obvious natural cause promising an easy solution of the question, and which will, probably, be found to depend on the character of the superficial deposits of the Doab, in these localities.

We may be very often deceived by the surface of plains, composed like those of Bengal and the N.W. Provinces, especially when we overlook the characters of subsoil, upon which the salubrity of localities depend.

In one tract, the dense retentive clays may be thrown up almost to the surface; in others, they may be buried to a greater depth. They may undulate and form basins, the hollows of which, being filled with light sediments, may give a treacherous character to the even surface, arising from concealed repositories of pent-up waters, which have no way of escape but by exhalation from the soil.

Such localities as these are further modified, and made independent of rains, to a certain extent, when situated on the banks of rivers where, during high floods, the waters rise above the dense clay forming their banks, escape unseen through percolating channels beneath the soil, and saturate without inundating the interior, as in the case of intermitting tanks in Bengal, which rise and fall with tides, without having any open communication with creeks or rivers.

At Prome, in 1853, there were five or six funerals daily, out of a European force of little more than a thousand, until the men were changed to the heights.* The unhealthiness of the low ground was caused, not by the overflowing of the river, but by saturation of sediments formed in a basin in clay, into which the river found its way through the porous soil, and which could not be drained.

Something of the same sort may account for the unhealthiness of Cawnpore, which might be difficult to rectify by drainage, as the water might still lodge beneath the level of the drains, and produce its effects

* This great mortality was checked by the change, although the hills to which the troops were removed were not more than 100 to 150 feet in height.

on the local climate just the same. Thus there may be no remedy for an ill-chosen site.

The very high ratio of deaths at Dinapore may be due to the same or similar causes. It would not seem to be the fault of the barracks, because the same high ratio of deaths occurred at Bhagulpore, in the same neighbourhood, before it was abandoned. Indeed, the most fatal group of stations in India for European troops, are those extending from Berhampore to Benares, which, with different kinds of barracks and other accommodation, present the same high ratio of deaths.

How are we to account for the fluctuation of mortality at Ghazipore, as shown by the returns for 14 years prior to 1836 to present a ratio of only 41·05 deaths per 1,000, while, for five years subsequently to 1836, until it was abandoned, the deaths amounted to 91·44 per 1,000?

This uncertainty of stations in the plains, even after heavy outlay for their improvement, would in itself be most discouraging, did we not know how inconsiderately enormous sums are spent without any proper guarantee or security for the health and interest of the troops —such matters being left to the management and control of the officer in charge of public works, who erects barracks and selects stations without responsible advice or control. Still, it may be doubted whether, under the best system, the results, as regards the health of troops, would be materially affected so long as they are located in the climate of the plains.

On the other hand, it is obvious, from the little

experience we have had of hill stations, as shown in preceding tables, that the advantages they present over the stations in the plains are so decided, that their general adoption would reduce the ratio of deaths in Bengal to something like that of the other Presidencies. "Returns, extending over a great many years," Sir A. Tulloch observes, "show that at Bengalore, as in the Nielgherry hills, and at Poonah and the Mahableshwar hills, European troops sustain no greater loss than 3 per cent. annually; that in the Bengal Presidency we have been less fortunate, probably because strategic may have hitherto overcome sanitary considerations." The same authority observes, with reference to Bengal, where both strategic and sanitary considerations have failed, "that the selection of healthier stations for our troops than those they have hitherto occupied is no longer a matter of choice, but one of necessity."

Range and Influence of Remittent Fever.—From all that is known on the subject of malaria, we may conclude, without pushing the question to any great nicety, that the cause lies between the soil and its properties with respect to water. Whether the retentive quality of the soil in itself is sufficient, under the influence of high temperature, to produce malaria, with or without organic mixture. Whether organic mixture with the soil affects its character in relation to malaria mechanically, in communicating to it retentive or absorbent qualities, or chemically, in affording material for the emanation of noxious exhalations, are points regarding which further observations are required. But high temperature, which seems to be essential in either case, would appear to act in two ways, *i. e.*, in disengaging noxious exhalations, and at the same time, in predisposing the body to their effects.

Free drainage is the most direct means of counteracting the influence of malaria; and this is naturally attained, more particularly in isolated mountain elevations: such situations are usually free from fevers, although these maladies are not uncommon in warm mountain valleys;* thus proving that it is only in proportion as mountains are better drained and ventilated than plains, that they are more free from fever.

For these reasons, I entirely agree with Dr. Martin in his evidence, that elevations of 2,500 to 4,000 feet would be sufficient to overtop malaria in Bengal and the N. W. Provinces.

Whatever the nature of the soil might be at such elevations, there would be such facilities for local drainage as to render the invasion of endemic remittents out of the question. Besides which, the lower temperature of such elevations would lessen the disengagement of malaria, render it more feeble, and the European constitution, at the same time, less predisposed to its effects.

Towns, barracks, and cantonments may be drained, with more or less facility, according to the elevation of their sites above adjacent rivers and plains.

A perfect system of drainage, to a degree that would render cantonments in the plains of India alto-

* *See* pp. 2 and 7. Shore Valley, in Kemaon, at an elevation of 4,500 feet, is remarkable for the prevalence of intermittent fever.

gether exempt from malaria, would be impossible, although the evil might perhaps be much lessened; still, the malaria could not be shut out from the undrained neighbourhood, although a judicious selection of trees adapted to the place, might assist as a barrier to the invasion of malaria from without.

The prevalence of fever in military cantonments in India depends on other causes besides malaria, such as exposure to the sun, functional derangement, and intemperance. If, however, the soldier be protected from malaria, the fevers resulting from other causes would be slight and manageable, both in the means of prevention and cure; because fevers, however contracted, in malarious districts have a tendency to assume the endemic, or severe type.

The extent to which the army is crippled by fever in the plains of India has suggested the proposal, strongly urged on the attention of the authorities by Dr. Martin, of locating the European troops in hill stations, raised above the influence of malaria.

The hill stations hitherto resorted to in the Himalaya and other parts, while they raise the European from the range of malaria, appear to expose him to other evils. Dr. Martin, in his evidence before the Committee for the reorganization of the Indian Army, therefore proposes to appropriate the lower medium ranges of hills throughout India for the permanent cantonment of our European troops, in lieu of the plains, and has adduced many powerful reasons to show the State necessity for such an arrangement.

In the West Indies, it is observed that an elevation of 2,500 feet has been found sufficient to remove the

European from the range of yellow fever. In consequence of this change, the mortality of our troops in Jamaica is now found to be little, if at all in excess of those serving in the United Kingdom. Although the West Indies is an insular climate, and may perhaps on that account, show more favourable results than might be expected to arise from a corresponding change in India, still, from all that we know on the subject, there appears to be no reason why we may not obtain as favourable results in regard to the remittent fever of India.

Hitherto it has been considered necessary that our European troops in India should be stationed near the great masses of native population in the plains, without reference to the salubrity of such sites. Hence it appears, from the evidence elicited before the Committee for the reorganization of the Indian Army, that from 1815 to 1855 the mortality, exclusive of casualties, amounted to 100,000 men—"the greater portion of whose lives might have been preserved had better localities been selected for the military occupation of that country."

Surely the question requires no stronger argument than this to force it upon the most serious and careful consideration at least of every medical officer in particular, whose duty it is to ponder over it, in order to discover by what practicable arrangement the selection of sites might be improved in the plains; and secondly, to point out what elevated spots may be available, free from the objections referred to, and otherwise adapted for the purposes of military cantonments.

With regard to the first of these questions, I have shown that the hottest portions of the plains of Bengal and the N. W. Provinces have been hitherto chosen, or rather pitched upon, for European stations, while other climates and localities in the plains are pointed out and generally referred to, which promise better in regard to climate.

It now remains to show, as far as we appear to have any information, how far we have the command of elevation in military positions that would place our European troops above the influence of the malaria of the plains.

Hill Climates for European Troops.—The want of broad elevated table-land in Kemaon, and the numerous deep valleys extending in every direction, must at once appear as an objection to that province for purposes of European colonization, or the location of European troops. These objections apply more or less to most parts of the Himalaya with which we are acquainted, to say nothing of the difficulty of access with the plains, as well as the other objections raised by Dr. Martin to these stations generally, that while they are elevated above the influence of fever, they are unfavourable for other disorders scarcely less destructive to our troops.

The climate and elevation of the Khasyah mountains, lying between Lower Bengal and Assam, are more promising. The elevations there, more especially about the centre of the table-land, are broad and extensive downs, with favourable soil and plentiful supplies of water, with coal and other resources at hand, suited to the enterprise of Europeans, which seem to leave nothing to be wished for in regard to this locality, but greater facilities of communication, in order to render it all that could be desired for the object in view.

There can be no doubt that any elevation on a rocky light soil above a thousand feet would be preferable to the plains. Although the heat during the months of April, May, and June, would be little less during the day than that of the plains, the nights would be cool and more refreshing; and the months of August, September, and October, would present a much lower temperature than the plains, as in the case of Hazareebagh, at an elevation of 1,500 feet.

If Hazareebagh, or any other part of the table-land on which it stands, were selected for the advantage it possesses of a lower temperature during the nights at all seasons, as well as both day and night during the three unhealthy hot months of August, September, and October—so fatal to our European troops in the plains—one hundred and twenty miles of railway from the heights to the Ganges at Monghyr, would make a large force stationed on the table-lands available for all duties now required of the European troops stationed at Dinapore, Berhampore, and Ghazipore.

Nor would the use of such a railway be necessarily limited only to military purposes alone, but might be made the means of furnishing cheap supplies of coal to the Ganges from the upper beds in the Damooda valley and its tributaries, as well as from the coal field of Curhurbalee in Purgunnah Coruckdyah, through which the line would almost pass.

Proceeding from Hazareebagh still further to the westward, passing over the table-lands of Sirgoodjah and Main-pat, of which we know little, we have a range of high country running parallel with the Ganges, presenting fine table-lands at Sohagepore, the elevation of which has been variously stated at from 3,500 to 5,000 feet. I believe the latter to be the most correct. This table-land is within a range of 130 to 150 miles, or less, in a direct line, of Mirzapore and Allahabad Here, as we have also excellent coal, we should have the same interest in making it available on the higher portion of the Ganges, where it would be still more valuable. But that is a small object compared with the climate this table-land would afford to our European troops.

The following is from Thornton's *Gazetteer** :—"Amarakantak, in the district of Ramgurh, Saugur and Nerbudda territories, lat. 22° 40′ N., long. 81° 50′ E. According to an account of some intelligent Brahmins, who had visited the temple, it is situated in the midst of the table-lands. The spot formerly the subject of conflicting claims, was adjudged within British territory in 1826, by the treaty of Nagpore. The place is one of considerable resort for Hindoo pilgrims, though the surrounding country is in general a wild and nearly pathless jungle, regarded by the natives as the abode only of wild beasts, demons, and savage goonds. Although only 120 miles south-east of the British station of Jubbulpore, it has been so imperfectly explored, that no tolerable approximation has been made to show its elevation above the

* Published under Authority: London, 1854.

sea—some making it 5,000, and others 3,500. The height, however, is sufficient to render the climate much cooler than the country about Jubbulpore, which has an elevation of 1,500 feet, and in the hottest time seldom exceeds 95° Fahr."

The annexed register of the thermometer, kept by a native at Amarakantak, and printed in the *Bengal and Agra Gazetteer*, 1842, shows the mean temperature of the month of May to be 19° Fahr. less than that of Allahabad.

	Amarakantak, May, 1838.			Amarakantak, May, 1838.	
	6 a.m.	Noon.		6 a.m.	Noon.
May 1	70°	88°	May 16	76°	87°
„ 2	64	89	„ 17	76	89
„ 3	74	90	„ 18	76	86
„ 4	78	91	„ 19	75	86
„ 5	78	94	„ 20	71	89
„ 6	75	86	„ 21	79	81
„ 7	71	84	„ 22	80	92
„ 8	68	79	„ 23	75	94
„ 9	66	82	„ 24	78	93
„ 10	62	84	„ 25	68	94
„ 11	72	88	„ 26	68	94
„ 12	72	88	„ 27	74	95
„ 13	62	89	„ 28	76	30
„ 14	72	92	„ 29	82	93
„ 15	75	86	„ 30	77	91

Mean, 73° at 6 a.m., and 89° at noon.

A writer in the work referred to, who visited the place in 1841, describes it as too cold for the usual winter crops of Hindostan, which are nipped by the frost.

During the month of May the climate is truly delightful. Everything around is fresh and green; the air is elastic and buoyant, with dew falling every night. Fifteen or twenty days seldom pass, even in the dry season, without showers.

On the east and south the table-lands at Amarakantak terminate by precipitous cliffs. On the north and west are broad elevated plains extending for 50 or 60 miles in the direction of north and west, over which the prevailing wind blows.

Captain Franklin, who visited Sohagepore on these table-lands, in 1821, speaks in high terms of the climate.

He found coal a few miles north-west of Sohagepore, and about an equal distance from the upper tributaries of the Sone river, and also at the confluence of the Tipan with the Sone river, about 30 miles from the source of the latter and the same distance south-east of Sohagepore.

Thus we appear to have elevated table-lands and plains, presenting fine temperate climates and abounding in natural resources, within 150 miles of Allahabad, but of which we have very little information —the tract of country never having been surveyed and rarely visited by Europeans.

Patchmaree.—About 150 miles to the eastward of the table-lands of Sohagepore, in the Mahadee hills, between Baitool and Hoshingabad, is the table-land of Patchmaree, situated in the Nagpore district. It is about five or six miles in diameter, composed of sandstone, and elevated, according to Captain Franklin, 4,500 feet. The soil is light and sandy. The usual cold-weather crops are grown here with difficulty, as the frost generally nips the buds, although it is not quite so cold as it is at Sohagepore. Ferns are said to be abundant, which indicates a moist climate.

	PATCHMAREE, 1839.			
	Morning.	Afternoon.	Sunset.	
May 12.	79	86½	88	Tattees, previously in use, here dispensed with.
„ 13..	80	86	88	Cloudy, with high winds.
„ 14..	77	93	88	Westerly winds.
„ 15..	76	93	90	Close.
„ 16..	79	91	84	Clouds, wind, and thunder.
„ 17..	76	90	77	Clouds and rain.
„ 18..	76	92	86	Clouds.
„ 19..	78	93	89	
„ 20.	79	95	90½	
„ 21..	76	91	85	
„ 22 .	74	88	84	Rain and thunder.
„ 23..	72	85	81½	Heavy rain and thunder.
„ 24..	72	87	78	Ditto, ditto, with wind.
„ 25..	74	87	76	Thunder and rain.
„ 26..	73	86	75	Clouds and thunder.
„ 27..	71	81	81	Cloudy, clear in afternoon.
„ 28..	72	87	85	Clear.
„ 29..	77	88	87	Clear.
„ 30..	73	88	84	Rain, clear forenoon.
„ 31..	76	85	80	Thunder, showery.
June 1..	70	85	80	Thunder and clouds.
„ 2..	71	84	78	Ditto and showers.
	76	88	84 =	Mean 82°·6 Fahr

Speaking of the climate, the writer, in the *Bengal and Agra Gazetteer*, from which the above register is taken, observes, that the natives speak favourably of it; certainly, in the month of May, the climate has been found to be such as to suit the most delicate European constitution, while the cold season is intensely cold and invigorating. It may also be mentioned that at Multai, above Baitool, the climate, although not nearly so cold as at Patchmaree, is perceptibly cooler than it is at Baitool.

Alluding to this table-land, but more especially to that of Amarakantak, already described, the writer alluded to, in the *Bengal Gazetteer*, concludes his remarks by the following observations:—"A bracing climate, mineral treasures of unexplored value, rich and unappropriated lands, abundance of moisture for all agricultural purposes, beautiful scenery, and a vast variety of products of almost every description, are to be found in these table-lands, to which Providence has been so lavish in its gifts, but which man has not yet been taught to appreciate."

Why should we have so little information regarding these table-lands, their climates and resources?

The truth is, hill stations have hitherto been looked upon rather with disfavour on the part of Government, as so many inducements for leave of absence to officers whose duties lay in the plains; but the case is altered when it is proposed to make them the permanent stations of European troops.

With the facilities that railways afford, there is no reason why many civil establishments, and even the seat of Government itself, should not be placed on the table-lands referred to, for which there would be ample space.

The moral effect of a large European force in such a position as Sohagepore would be much greater than that of the same force distributed widely over the country, while troops might be readily thrown on any part of the low country in which their presence would be likely to be required, by means of railways.

On the other hand, by following the example of the natives of India in the adoption of their sites in the plains for all our great establishments, we lose sight not only of the sanitary wants of European troops in India, but also of the first principles of our own superior civilisation, which should teach us the value of the most salubrious spots, at least for our own institutions and our own communities.